Moment of Grace

Moment of Grace

The American City in the 1950s

Michael Johns

UNIVERSITY OF CALIFORNIA PRESS

Berkeley / Los Angeles / London

University of California Press
Berkeley and Los Angeles, California

University of California Press, Ltd.
London, England

© 2003 by The Regents of the University of California

Library of Congress Cataloging-in-Publication Data

Johns, Michael, 1958–.
  Moment of grace : the American city in the 1950s / Michael Johns.
      p.   cm.
  Includes index.
  ISBN 0-520-23435-9 (alk. paper)
    1. Cities and towns—United States—History—20th century.
  2. Metropolitan areas—United States—History—20th century.   I. Title.
  HT123 .J59   2003
  307.76'4'09730904—dc21   2002016564

Manufactured in the United States of America

12   11   10   09   08   07   06   05   04   03

10   9   8   7   6   5   4   3   2   1

For my mother and father

Contents

Illustrations

CHAPTER I

The City at Its Peak

America reached its peak as an urban society in the 1950s. This was the moment when the American city of factories, downtown shopping, and well-defined neighborhoods, vitalized by a culture of urbane song, dress, and manners, achieved its consummate expression; when new suburbs relied on cities for jobs and manufactured goods; and when the residents of those suburbs epitomized the nearly dogmatic optimism of the time and belonged to a dense network of groups and associations. But this was also the moment when all that solidity, urbanity, and community began to fragment, giving way, eventually, to the very different cities and suburbs of today.

Consider the typical downtown of the '50s. Most of the buildings were constructed around the turn of the nineteenth century, some fine art-deco structures had been built in the 1920s, and a few modern towers were just going up. Together they blended brick, stone, and steel into pleasing patterns that formed a lively retail and office district. The longest, sleekest, most lavish cars ever made cruised the streets. Walking those streets was a people whose sophistication relied more on formality than refinement, but who nonetheless gave American cities what writer Jan Morris described as their "moment of grace." That moment called for a jacket and tie, gloves and high heels, the manly etiquette of pulling back the lady's chair, lighting her cigarette, and paying the bill. The buildings, the business, the formality: no wonder historian James Vance looked back on his 1954 study of American downtowns and said, "If we had done that study much later, even five years later, it would not have been as good. We would have had to say the downtown was past its prime, that it was once better."[1]

The essential parts of a downtown—the rail station, trolley lines, office buildings, grand hotels, department stores, five-and-dimes, theaters, city hall, a nearby warehouse district, and a wholesale produce market—had been in place for fifty, sixty, or, in some cities, seventy years. By the late 1950s, however, downtown was becoming something else. Streetcar rails were ripped from the ground. An elevated highway sliced through the warehouse sector. Glass skyscrapers replaced brick buildings. Fewer people went downtown to shop, see a play, or meet a train. "Downtown is in transition," warned the *Saturday Evening Post* in 1960, "and a new pattern of urban development is emerging. It is a pattern fundamentally different from anything the city dweller has known before."[2]

The magazine was right. The 1950s were the last time the city dweller regularly visited downtown. Mayors would no longer refer to downtown as "the hub and nerve" of the metropolis. Never again would downtown embody the industrial age in a reassuring mass of brick and terra cotta, in detailed cornices and cast-iron lamps, in trains and trolleys, and, as a final touch, in the steely arrogance of the new cars and skyscrapers. By 1960 nearly everyone conceded that "the great magnet of the city has changed its polarity. Where once it attracted, it now repels."[3]

Since then, mayors, city planners, and redevelopment agencies have been trying to make their downtowns attractive again. In recent years, they have had some success. Many cities are installing old-style street lamps, and a few have brought back streetcars from the '40s and '50s. Big retailers hang old-fashioned signs on their downtown storefronts. Fine restaurants imitate art-nouveau interiors of a century ago. New diners display '50s decor while serving red-snapper sandwiches with spicy mango sauce. Young professionals move into converted warehouses. Restored hotel bars feature cigars, martinis, and crooners singing "Summertime" and "My Funny Valentine." The newest hotels, office buildings, and downtown baseball stadiums mimic earlier facades, an architectural plagiarism that reveals a lack of ideas but improves upon the designs of the '60s, '70s, and early '80s.

None of that means downtown is likely to regain its former charm or its once great economic and cultural power. Today's downtown is primarily an office and hotel center, just one of several in the metropolitan region. Fewer people shop, live, or seek entertainment downtown. The city's economy is more efficient than it was fifty years ago, but its workings are not as obvious as when docks, rail yards, factories, warehouses, and produce markets occupied the edges of the central business district. The urban economy, moreover, has lost much of its local flavor. Many companies that got started in a city, and had their offices downtown in the '50s, have since then moved to suburbs, gone out of business, or been taken over by firms from elsewhere. And if the average city dweller appreciates the new clusters of bars, clubs, and restaurants, and applauds the attempt to restore some urbanity to downtown, he also knows that the setting looks like a stage set made out of images and things from other times and places. Gone are the days, remembers an old man about Tucson in the '50s, when "everything was—that's it, Downtown."[4]

Downtown united a city whose neighborhoods, wrote E. B. White in 1949, were each "a city within a city within a city." Many families had lived in their neighborhoods for decades and had relatives a few blocks away. Housewives and children filled stoops and porches during the day. Residents walked to a nearby shopping street that offered, according to White, just about everything:

a grocery store, a barbershop, a newsstand and shoeshine shack, an ice-coal-and-wood cellar (where you write your order on a pad outside as you walk by), a dry cleaner, a laundry, a delicatessen (beer and sandwiches delivered at any hour to your door), a flower shop, an undertaker's parlor, a movie house, a radio-repair shop, a stationer, a haberdasher, a tailor, a drugstore, a garage, a tearoom, a saloon, a hardware store, a liquor store, a shoe-repair shop.[5]

Retired people remember a stable and close-knit neighborhood culture. They reminisce about good factory jobs, local union halls, and prosperous retail streets. They also recall the ethnic and sometimes racial prejudices that typically formed an important part of a white community's sense of itself.

Black neighborhoods were poor, segregated, and crowded with newcomers from the South. Yet they, too, were cities in miniature, filled with doctors and laborers, homeowners and newcomers, the lettered and the illiterate, who all lived, shopped, and enjoyed themselves on the same streets. Optimism softened the hard edge of poverty and prejudice. "There was a wonderful vivacity and excitement," recalled a former resident of Harlem. "I do think that those of us who were growing up then held that view of life in the United States. There was a sense that if you kept your nose clean, and if you went to school, and you held a good job, and you made a little money, and you washed and ironed your clothes—that it was all going to turn out all right."[6]

But it did not turn out right for many people, white and black, who stayed in their neighborhoods during the '60s and '70s. In 1955, a magazine could still write a story about "Worcester at work" as a Massachusetts city that manufactured goods as well as civic pride. Worcester's story, the article concluded, "is the story of the growth of industrial America. Such communities made this country what it is today." Similarly, it made sense in the mid-'50s to paint "Trenton Makes, the World Takes" on a railroad bridge and for a newspaper headline to proclaim, "Great Industrial Variety Helps Philadelphia Stay a Solid City in Boom." Yet manufacturing growth was coming to a close in cities like Worcester, Trenton, and Philadelphia—and in Detroit, Chicago, and San Francisco as well. By 1960, many manufacturers had scaled back or moved to suburbs; trucks hauled freight that had required rail transport just a few years earlier; and docks berthed container ships and laid off longshoremen.[7] Just when white neighborhoods started to lose jobs in urban manufacturing, warehousing, and ship and rail transport, they also began to feel the loss of young couples who had been moving to suburbs since the late '40s. Most black neighborhoods degenerated rapidly in the early '60s. They lost access to blue-collar jobs, they lost block after block to urban renewal, and they lost their middle- and upper-class residents to better neighborhoods and suburbs.

That is not to say the condition of every city neighborhood worsened after the '50s. Many kept their retail streets intact, even though supermarkets replaced grocers and butchers, big drugstores supplanted pharmacies and five-and-dimes, and television and multiplex cinemas eventually forced most neighborhood movie theaters to close. Furthermore, some of the neighborhoods that did lose residents and stores during the '60s and '70s have been rejuvenated in recent years. Hispanic and Asian immigrants now shop on commercial avenues lined with small stores run by merchants who extend credit to customers and sell food and drink from the old country. Young urban professionals, meanwhile, continue to transform distressed areas into fashionable neighborhoods, centered on retail streets that feature fancy supermarkets, art galleries, health clubs, restaurants, wine stores, sports bars, real-estate brokers, hair-styling salons, dessert bakeries, antique shops, and chain stores selling clothes, drugs, bagels, coffee, videos, and vitamins. For all their virtues, however, neither the immigrant community nor the yuppie district is the world in miniature that the average neighborhood was in the '50s. That stable little world may have been insular, and even stifling. It certainly afforded fewer of what now are called lifestyle choices. But it did provide, according to its residents, a "center of gravity," a "sense of continuity," and a "close familiar whole."[8] Such images describe few neighborhoods today.

Just as the downtowns and neighborhoods hit their peaks, a great "surge to the suburbs" initiated a profound shift in the economic and cultural geography of the American metropolis. That surge, wrote the *Saturday Evening Post,* was "motivated by emotions as strong and deep as those which sent the pioneer wagons rolling westward a century ago."[9] Young parents built vigorous communities to replace the relatives and institutions they left behind in the old neighborhoods. They exuded confidence, too, thanks to rising wages, an unprecedented array of affordable goods, the conviction that their kids would do even better than they themselves were doing, and the prospect of selling their houses someday and buying better ones in newer suburbs.

At first the people of the new suburbs relied on their nearby cities. Many suburban residents worked in city factories and office buildings and traveled to cities on weekends to visit parents or siblings in the old neighborhoods. Some even retained a semblance of urbanity, which usually meant dressing up to drink highballs in each other's living rooms or an occasional night out at a downtown club. By the early '60s, however, the average suburb had become independent of its city. Few suburbanites worked, shopped, or looked for entertainment downtown. Most of them had given up any sense of urban decorum and reserve in favor of convenience and informality. And they were buying many more goods manufactured in suburban plants and supplied by suburban wholesalers.

It is no surprise, then, that one must go back to the '50s—when America hit its peak as an urban society—to find a way of life that looks really different from today's. For the '50s mark the end of widespread urbanity, strong neighborhood identities, downtowns that unified their cities, a national economy powered by urban manufacturing districts, and suburbs still attached to and dependent upon cities. The '50s were the last time, moreover, when the society recognized as natural and proper such cultural traits as clearly defined roles for men and women, the legal separation of blacks and whites, and the expectation that teenagers knew their places in an adult world.

If the '50s seem distant because they mark the end of so many things now considered traditional, old-fashioned, or even backward, they also seem far away because of a feature quite particular to those years: an overall cultural coherence. That coherence was manifest in a nearly blind faith in progress, a strong sense of patriotism, a forward-looking attitude that lacked nostalgia, a prevailing dress code, a homogeneous mass market for consumer goods, and a process of assimilation so absorbing that writer Philip Roth remembers it as a period of "fierce Americanization."[10] The coherence of the '50s looks especially foreign from the perspective of to-day's society, which affords nearly unlimited personal freedoms, places lit-tle trust in government, and relies heavily on the past for ideas, images, and styles.

Just as a peak implies an end point, so too can an end point mean a new beginning. Such was the case with the '50s. For, in addition to consum-mating an older, urban way of life, those years also established the frame-work for today's way of life. The teenagers who now are such a huge mar-ket for popular culture can trace their origins as a separate group to the '50s. The struggle for equal treatment by blacks in northern cities created the modern era of race relations. Television, long-playing records, and all manner of new household appliances initiated the age of consumer elec-tronics. And it is easy to see, in the postwar suburbs especially, the origins of today's casual manners, shopping malls, dependence on cars, greater metropolitan areas comprised of one subdivision after another, and a sense of community that relies more on voluntary associations and chil-dren's activities than it does on religion, ethnicity, or place.

That America hit its peak as an urban society in the 1950s explains, in large measure, why those years are a turning point in the life of the na-tion, why they have such a distinctive character, and why people find them so intriguing today. It seems fitting, therefore, to examine the 1950s by looking at the downtowns, the neighborhoods, and the suburbs of that decade's cities.

CHAPTER 2

The Downtown

Downtown "is so familiar to the average citizen that he is likely to take it for granted," wrote two urban geographers in 1954.[1] It was easy to take for granted because almost everyone who lived in a city went downtown regularly—to work, shop, see a movie, hear music at a nightclub, or call upon a lawyer or a dentist. Even those who did not live in cities were familiar with downtown's commanding presence. Magazines displayed photographs of soaring art-deco skyscrapers and the glass and steel towers that rose up after the war. Tall buildings, along with rail stations, hotel lobbies, and sidewalks, were standard settings for movie scenes in the '40s and '50s. Everyone knew that theaters, clubs, and department stores were places of culture and sophistication. The average citizen also knew that most of the nation's business was conducted from the downtown offices of big cities, and that most of its corruption took place along nearby docks and rail yards. What citizens did not know in 1954 was that they would never again be able to take their downtowns for granted.

The Central Business District

Docks, rail yards, and warehouses that had sat empty during the Depression were loaded during the Second World War and then were filled with the nation's goods for the next fifteen years. Retail stores lured customers back after the war, while office buildings regained tenants. Right through the '50s, concludes an urban historian, downtown was "still very truly a central business district."[2]

OFFICE WORKERS

The business district's main function was to serve as the office center of the metropolis. Almost 95 percent of all office space was still downtown in the '50s, and nearly every square foot of it was occupied. Big companies had their home offices in the central business districts of New York, Boston, Chicago, Philadelphia, Pittsburgh, and San Francisco. Thousands of smaller businesses occupied office buildings in cities like St. Louis, Milwaukee, Kansas City, and Baltimore. Even the central business districts of Oakland, Tucson, and Worcester contained the offices of local banks, manufacturers, wholesalers, trucking and shipping operators, law and accounting firms, advertising and insurance companies, and doctors, dentists, and private investigators. Rapid economic growth after the war resulted in more businesses and more employees. An expanding corporate sector created a bigger bureaucracy. A steep rise in industrial efficiency meant fewer workers made more goods, while more workers advertised, counted, and sold those goods. The upshot was a tremendous growth in office work of every kind. By the middle of the '50s, the number of white-collar workers—accountants, lawyers, secretaries, salesmen, midlevel managers, and paper pushers of all sorts—had for the first time surpassed the number of blue-collar workers.[3]

Nearly every man in an office district, from young fathers with desk jobs to longtime owners of small businesses, dressed in a jacket and tie, and if he was over thirty, he wore a hat. Corporate businessmen, and professionals like company lawyers and advertising men, set the standard. They wore charcoal-gray suits with striped ties, polished loafers, and pink, blue, or white shirts.[4] They did not sport beards, sideburns, or mustaches. But many did use Vitalis, or some other hair lotion or cream, for what the advertisements called a just-combed, well-groomed, or under-control look. Their hair was trimmed above the ears, cut straight across the nape, and neatly combed back or parted on the side.

This sharp look reflected a business culture that, after fifteen difficult years, once again had money and confidence. Many of the younger businessmen and professionals, remembered writer Norman Podhoretz, actively "cultivated an interest in food, clothes, furniture, manners—these being elements of the 'richness' of life that the generation of the 30's had deprived itself of." The rapid growth of large companies after the war, and their congenial relationships with labor unions and the federal government throughout the '50s, gave corporate culture the size and certitude it needed to establish a dominant protocol. "Some critics," said steel magnate Ben Fairless in 1956, "deplore the fact that all businessmen today seem to be modest and polite duplicates of one another instead of flamboyant 'characters' as in the old days. But you cannot have it both ways, and today's Man in the Gray Flannel Suit is ten times more efficient than his more colorful, more hot-tempered predecessor."[5]

William Whyte went so far as to rebuke the city's white-collar worker for his caution and conformity. But Whyte also acknowledged that "society is not out of joint for him, and if he acquiesces it is not out of fear that he does so. He does not want to rebel against the status quo because he really likes it—and his elders, it might be added, are not suggesting anything bold and new to rebel *for*."[6] A spate of novels, academic treatises, and popular books—good examples are Cameron Hawley's *Executive Suite* and *Cash McCall;* C. Wright Mills's *White Collar* and *The Power Elite;* Vance Packard's *The Hidden Persuaders;* and Whyte's *The Organization Man*—questioned the expanding power of corporate culture and the groomed look and neat conformity of its downtown workers. Those writings registered a new reality: corporations grew in number, size, and strength, while more company heads had started out as salaried employees and fewer executives owned their firms. It was also true that the number of small-business owners fell as the ranks of company employees swelled. So it is no wonder that the business culture of the '50s was America's most formal, orderly, and coherent.[7] While that culture was best expressed in Manhattan, Chicago, and San Francisco, it set the tone in downtown Denver and Cincinnati as well.

Movies like *Woman's World* and *Executive Suite,* live television dramas like *Patterns,* and countless novels with titles like *Front Office* all portrayed a business culture that revolved around men but relied increasingly on women. Female office workers figured prominently in movies, photographs, and novels because those women had a big effect on the culture of business and the look of downtown. The number of secretaries and receptionists doubled during the 1950s, and downtown streets were heard to "echo to the click-click" of high heels worn by stenographers, bookkeepers, ticker operators, file clerks, messengers, and pages.[8] The efficient, intelligent, and highly respected executive secretary had the top job among women office workers. She was over thirty, put her hair up in a bun, was as "neat as a trained nurse," and organized much of her boss's daily business. At the bottom were young women who tapped all day on new electric typewriters. In between were bookkeepers, file clerks, and copy editors. While some of these women had gone back to work after sending their youngest children to school, most of them worked during the years that bridged high school or college and marriage and mother-hood. "What kind of job or occupation do you think offers a young woman the best chances of finding a husband?" asked Gallup. A third said their best bet was a downtown office job.[9]

The number of women who worked downtown as managers, account-ants, editors, lawyers, and reporters also increased during the 1950s; but the ratio of women to men holding skilled positions in the professional and business world declined between 1940 and 1956. Most women who did well in business worked for themselves as owners of restaurants, clothing stores, and gift shops. Of the million so-called executives in the United States, at most five thousand were women. Most of these were self-employed or made their way up through small- and medium-sized firms. "Officers of big companies," reported *Fortune* magazine in 1956, "are some-what embarrassed to admit the scarcity or lack of openings for women."[10]

The 1959 movie *The Best of Everything* portrayed the life of a corporate woman. Joan Crawford plays a gruff editor in a large publishing house. She faces competition from a much younger woman, played by Hope Lange, who is climbing the ladder from typist to copy editor to editor. Lange's character is generous and affable. She may not need (or has yet to develop) the hard edge of Crawford's character, who is, after all, part of the first generation of women to break into the corporate world. But Lange still faces a business culture that does not check men from patting a woman on the behind, commenting on her body, or making unwel-come advances in the office. In a scene from the movie version of Sloan Wilson's novel *The Man in the Gray Flannel Suit*, Gregory Peck gets in-troduced to his secretary. "They always give a new man the prettiest sec-retary," Peck's boss tells him. "Keeps 'em happy during the breaking-in pe-riod." It was the rare secretary who balked at doing menial favors for her boss. A stern warning accompanied an excellent recommendation for one such secretary, who was portrayed in Jerome Weidman's novel *The Enemy Camp:* "But if you ask her to send down to the drug store for a container of coffee, I warn you, Mr. Hurst, when she brings it into your office Miss Akst will make you feel they sweetened it not with sugar but with a cou-ple of teaspoons of ground glass."[11]

The rapid increase in the number of office women, the expanding power of large companies, and the smart look of white-collar men were all part of a central business district at its peak. The office sector may have been the most important part of that district, but shopping was its most conspicuous. "The citywatcher," a journalist wrote of downtown San Francisco, who avails himself of one of the thoughtfully placed benches on a shopping day will observe matrons archaic and matrons nubilic and matrons of resolute mien. . . . Prim high priestesses from the Junior League world of good works march, conspicuously white-gloved, into the furriers'. Models, overdressed and underfed, prance by like high-schooled horses. Stately clubwomen, boned and buttressed like ambulant churches, parade to the inaudible organ swell of "Pomp and Circumstance." Convoys of secretaries break cover for coffee. Clerks flit by hastily. Buyers stroll in pairs. The men are few.[12]

SHOPPERS

The women visited hat shops, record marts, bookstores, pharmacies, swank boutiques, and clothing outlets. But department stores and five-and-dimes were the two places in downtown most responsible for providing city dwellers with common shopping experiences, and they are the places most typically and fondly remembered today.

A writer in the early '50s could still describe the downtown department store as a "community institution" and one of the "great pillars" of the American economy because it was the city's prime place to shop and was thus visited regularly by most of the city's women. As the "mother store," it oversaw the establishment of suburban branches, dictating what to sell and where to advertise, making the branch manager, according to *Fortune* magazine, "a much less exalted character" than his counterpart downtown.[13]

As early as 1951, however, a student of retailing predicted that the "type of customer is going to be lower and lower as the years go on; and in twenty years . . . the downtown store will become a basement-and-budget type of operation only." He was basically right. For even though most city residents and many suburban dwellers still shopped in the city center—except in Los Angeles, where downtown accounted for just one-third of the metropolitan area's department store sales in 1950—what was true of Detroit in 1956 was true for other cities as well: "The downtown property owners," a magazine warned, "have been carrying on a consistent campaign to remind their old customers that shopping downtown is a pleasure, but they have had small success." The loss of customers to suburban shopping centers forced a few downtown department stores to close at the end of the decade. By the early '60s, the dependence of department stores on downtown was seen as a handicap. At the time that Manhattan's Saks Fifth Avenue closed its women's hat department in 1965, downtown department stores everywhere were losing not only their hats, but their shirts as well.[14]

(handwritten margin notes: "where did" "where the high class customer go?")

Even though business leveled off by the middle of the 1950s and the class of customer slipped a little, most downtown department stores held their own throughout the decade. "People still like to shop downtown," said a Gimbel's executive about Philadelphia in 1954. "If we give them attractive stores with good stocks, they will continue to shop downtown."[15] They did just that—for a few more years.

Everyone of a certain age remembers shopping regularly in big-name stores like Macy's, Filene's, and I. Magnin, or in smaller ones like Denholm's and W. T. Grant's. Susan Porter Benson, in her book about department stores, recalls shopping in downtown Pittsburgh during the late '50s:

> My mother introduced me to the wonders of the department store, shepherding me from store to store, teaching me more by example than by precept how to shop and rewarding good behavior with a ladylike lunch in one of the store restaurants. For many years, I thought that chicken à la king as served in Kaufmann's dining room was the summit of gourmet indulgence and genteel elegance.

Women from across the country shared their memories with her: "Whether they recall the heady metropolitan excitement of Macy's Herald Square store or the small-town gentility of Colorado Spring's Hibbard and Company, they all testify to the material and imaginative hold these giant stores once had on American womanhood." The big stores gained their hold on American women in the late 1800s and they lost it in the early 1960s.[16]

The department store offered attentive service, first-class goods, and an atmosphere of comfort. All of that had been expensive and had presumed a fairly well-to-do clientele until the late '40s and '50s, when many more working-class women had the money to shop in department stores and enjoy their service, ambiance, and formality. Saleswomen ran the floors, worked closely with shoppers, and treated even obnoxious customers with reserved courtesy. Elevator operators called out the departments on each floor. Store restaurants gave customers "a ladylike lunch." People of all persuasions got dressed up to shop. The department store, like the rest of downtown, was an adult world. Over 85 percent of those in downtown stores were adults shopping alone; in shopping centers, by contrast, only 43 percent of shoppers were by themselves and over a third had children in tow.[17]

If department stores were the big draw, all kinds of clothing stores, shoe stores, drugstores, and miscellaneous specialty stores also attracted shoppers downtown. The stores people remember most fondly, however, were the five-and-dimes—stores like Kress, Kresge, and, of course, Woolworth's, which thought of itself as "everybody's store."[18] The five-

and-dimes did, in fact, appeal to nearly everyone because they carried almost everything. Window displays featured dolls, scarves, plastic wallets, or shiny metal pots. Inclined counters held boxes of buttons, spools of thread, and bins of ribbon that you could have cut to size. There were sections full of unboxed lamps and lampshades, of metal towel hangers and soap dishes ready for inspection, of cups and saucers laid out on tables. Shelves were loaded with household gadgets, small hand tools, and every size and shape of screw, hook, and nail. And there were loads of costume jewelry. Each section of the store had a wooden booth enclosing metal cash registers where clerks rang up sales. Some still sent their money out via pneumatic chutes.

A five-and-dime's lunch counter took up a large part of one wall. Colorful signs in block print advertised "Swiss Steak for 65 cents"; "Hires Rootbeer—5 cents for Big Glass"; "Pumpkin Pie—10 cents." Unnatural images of food and drink were painted in lurid colors and at great size. Some of the food even tasted unnatural. A woman remembers tricolored ice cream sandwiches enclosed in square cookie covers that had "the flavor of pasteboard—a taste that became addictive."[19] The look and feel of the lunch counter, like so many other things at the time, was solid and substantial. Steel-rimmed glass containers displayed pies, doughnuts, and egg-salad sandwiches. Stoves, refrigerators, and drink dispensers, along with the big blenders of the soda fountain, all gleamed with polished metal. Formica covered the countertop, and leather or Naugahyde padded the round swivel seats set on metal tubes.[20]

Five-and-dimes and department stores belonged to a shopping district that served, according to a 1954 study of downtowns, "the entire community rather than any one part of the city." Blocks of apartment buildings, residential hotels, and houses still flanked sections of downtown. Their inhabitants, many of whom were poor or close to poor, used the bars, eateries, movie houses, laundries, pawn shops, delis, and grocery stores that were scattered along the fringes of the central business district. But most downtown shoppers and pleasure seekers came from all across the city. A study of shopping patterns showed that a quarter of city residents had shopped downtown within the previous week. They bought three-quarters of their clothing and half of their home furnishings in the central business district. They made a third of their visits to doctors and dentists in the city center. And they watched about a third of their movies in downtown theaters.[21] A lot of people went downtown for the mere enjoyment of the crowd. There was no place like downtown, in other words, for combining shopping with running errands, watching people, meeting friends, eating lunch, or seeing a movie.[22] It was easy to get to as well. The number of riders on public transit hit its peak in the late '40s, but throughout the '50s most people continued to get into and out of downtown on subways, streetcars, and buses.[23]

Public transit served downtown and downtown served the entire city. But not everyone was served equally. A woman recalls shopping in Cleveland:

We always got dressed up in clothes reserved for going to church to go shopping *downtown*. My patent leather shoes glowed, and my mother inspected my ears and my little white gloves to be sure they were clean. The most important thing, more than finding what we wanted, was to let the white folks know as soon as they saw us that we were okay—meaning we were clean enough to try on the clothes. Some of the sales ladies recognized my mother for what she was—a classy lady who could discuss Elizabethan literature with the best of them but right now was bent on getting most out of this sale. Other clerks would follow us around suspiciously, not speaking, not letting us out of their sight.[24]

Yet blacks did enjoy a certain security and independence in the downtowns of most northern cities. It was generally recognized that they had a legitimate place there as shoppers, strollers, and workers—if not as patrons in many hotels or as employees in some stores. Of Chicago in 1961, two black sociologists observed: "Negroes now move about in the central business district with a confidence that they have not shown since the Great Migration [of the 1920s] shattered the structure of the pre–World War I 'Golden Age' of race relations in Chicago."[25] It does seem as if blacks were freer to shop, work, and enjoy themselves in northern downtowns during the '50s than they had been in previous decades. Such freedom did not always include immunity from suspicious looks, whispers, and slurs, but it was a freedom that did not exist in the few big cities of the South. In 1960, for example, a man wanted to rest over a cup of tea while shopping with his fiancée in downtown Nashville. "It was such a normal thing to do," he said, "and then we realized it was impossible." To be able to do such a normal thing in the downtown of a southern city, blacks had to perform the extraordinary act of sitting-in at the places that refused to serve them.[26]

DOCKS, WAREHOUSES, AND RAIL YARDS

Downtown was more than the city's office center and its main place to shop. A few blocks from the skyscrapers and department stores was a rugged world of rail yards, small factories, brick warehouses, and, in port cities, docks. Most downtown shoppers and office workers avoided this part of the central business district. But they read about it in the newspapers, heard stories from people who worked there, and made occasional visits to an automobile-parts supplier or a furniture factory that sold directly to the public. Nearly everyone, moreover, had ridden a passenger train into the city. To take that ride, wrote an urban planner, was to see that "our water fronts, railway yards and industrial areas, with their dumps, automobile graveyards and general disorderliness, are often as disreputable as the worst residential slums."[27]

What may have looked disorderly to the average person was actually a vital place of storing, making, and moving that connected each downtown, and its entire city, to the rest of the nation. The United States was at its peak as a manufacturing economy in the 1950s, importing just 3 percent of its gross domestic product. Its myriad economic sectors and regional geographies created a nearly self-sufficient industrial machine. Its cities supplied one another, like never before or since, with raw materials, manufactured inputs, and finished products.[28] Although trucks were increasing their share of the haul and suburban wholesalers were capturing more of the business, the great bulk of merchandise was still carried into and out of cities by rail cars, barges, and freighters. Most of it was loaded and deposited at pier sheds, rail yards, warehouses, and small factories that in the majority of cities stood just a few blocks from department stores and office buildings.[29]

The workers who did all of the necessary lifting, loading, and hauling at these places wore jeans, overalls, cloth caps, and, in winter, heavy wool coats. They carried metal lunch pails packed with sandwiches or pieces of cold chicken wrapped in wax paper. Once in a while they ate at a diner, where they settled themselves into booths framed in chrome tubing and covered with Naugahyde. They rested their elbows and forearms on Formica table tops while their meatloaf was warming up behind the glistening stainless steel of the back bar.[30] A few of them walked to work. Some drove automobiles. Most took streetcars. Close to the ground and heavy with steel, the streetcar seemed to hug the surface of the street. The dull clang of its bell, the air hissing from its brakes, the entire steely mass streamlined into rounded sleekness—the streetcar, like the diner and the blue-collar man himself, was part of an industrial age then coming to an end. Buses and automobiles were replacing the streetcars and passenger trains. Trucks and planes were supplanting freight cars and cargo ships. Container ports were about to finish off the old docks and most of their longshoremen. Downtown warehouses and small factory buildings were being torn down for off-ramps, parking lots, and convention centers. Just like the department store, however, this old world of warehouses, rail yards, piers, small factories, and manual laborers remained an integral part of downtown throughout the '50s. It was maintained, for the moment, by the economic surge that would eventually render it obsolete.

The docks were the most mysterious and notorious part of this old industrial complex. New York's was the world's busiest harbor until 1960. Wharves, ships, and cranes jammed the waterfronts of Brooklyn, Hoboken, and Weehawken. Even the West Side of Manhattan, from its lower tip to West Fifty-ninth Street, had 110 piers, 24 ferry slips, and 8 float bridges for railroad freight cars. More passenger liners left the city in 1957 than in any other year; that was also the first year more Americans traveled to Europe by air than by sea.[31] Not only New York's but the docks of every port city—New Orleans, Seattle, Long Beach, Mobile, Oakland, Baltimore, Boston, Philadelphia, and Miami, as well as lake and river ports like Chicago, Memphis, and St. Louis—were crowded with activity. "There is nothing quite like a walk along the waterfront," a journalist wrote of San Francisco around 1960. "Cavernous pier sheds, mysterious when their doors are shut, may reveal imports from all the free world. Piggyback boxes wait by the sidewalk for the transport truck that will take them to Denver, Chicago and points east. . . . Sleek hulls loom great overhead. Fingerlift trucks scuttle, freight locomotives shuttle. Red stack tugs bob at anchor."[32] Those tugboats nuzzled ocean liners into their berths, bossed freighters around with their flattened snouts, and towed barges filled with scrap, gravel, and rail cars. They did so with a certain cheek, plowing the water with high prows and spewing bursts of thick black smoke when straining to push or pull a great load.

The docks, wrote a female observer, were "a man's world. Women may walk here, especially if escorted, or if they mind their own business, striding along and speaking to no one. . . . But they will be stared at openly, admired if pretty, and possibly whistled at. The feminine walker should know this." The '50s were the last time the docks swarmed with men. They transferred cargo to freight cars, hauled goods to nearby warehouses, and lowered laden pallets into holds. The men of the piers, wrote Paul Jacobs, a friend of labor, were

hard-bitten, rugged, and profane dock wallopers whose hand hooks hang down from the back pockets of their jeans as they descend into the cargo holds. And life on the 'front is still far from easy. The longshoremen get high pay, have good working conditions, and hire out by rotation from the union hall instead of by the corrupt shape-up, but they still get crushed in the 'tween-deck hold. Three-fingered hands are not at all uncommon, grasping heavy coffee mugs in the dock-side cafés.[33]

The shape-up was a shameful scene of grown men crowding around a hiring boss who picked the day's crew. Union pressure had abolished the shape-up by the middle of the '50s, but cutthroat shipping companies and brawling teamsters and longshoremen's unions still dominated the piers. The teamsters controlled most of the trucking and warehouse business and therefore much of the waterfront. On that basis they grew rich enough to build themselves a new headquarters in Washington that had a penthouse lounge, an executive suite, a movie house with almost five hundred seats, and a two-story garage with parking spaces deep enough for all nineteen feet of a Cadillac. Local teamsters' offices and longshoremen's halls were frequented by what Jacobs described as the "new middlemen of industrial relations," the "operators," that is, who parlayed personal contacts with union leaders into jobs as "management consultants" for trucking, shipping, and insurance companies. Some liked to "smoke enormous cigars, wear suede and calf shoes, silk shirts, and hand-painted ties." A few were wealthy enough to "carry fat rolls of bills in diamond money clips and drive always-new Cadillacs to expensive steak houses."[34]

A Philadelphia newspaper acknowledged that pilfering took place along its miles of docks, but saw no signs of organized racketeering.[35] Congressional hearings and grand-jury indictments in the early '50s cleaned up other piers as well. But there were still "sinister" ports where a third of the leaders in the International Longshoremen's Association had criminal records and took gratuities from employers in exchange for "goodwill"; where theft was organized, commonplace, and involved payoffs to police and politicians; and where union dues were not clearly accounted for. Kickbacks, theft, and loan-sharking, according to a journalist, made "decency still a stranger in the crime-caulked system behind the bulkhead sheds."[36] The city docks were still a world, in other words, for characters like Johnny Friendly, the local union boss of some Hoboken piers in Elia Kazan's film *On the Waterfront* (1954). Friendly said a lot when he said, "We got the fattest piers in the fattest harbor in the world. Everything moves in and out. We take our cut."

Much of what moved in and out was warehoused in two-, three-, and four-story brick buildings packed into an area between the waterfront and rail yards on one side and the office and shopping district on the other. The front of a building advertised the company's name and function—John Dais Wholesale, Meyer's Produce, Wilson Meats, Butler's Storage—in plain block letters. Wooden or canvas awnings sheltered loading docks. Trucks backed up to some; freight cars sidled along others. Pallets, boxes, and crates were strewn everywhere. Scattered throughout this warehouse district were a row of office-supply stores, a block or two of discount furniture marts, an intersection of auto-parts suppliers, and a cluster of low-quality housing, as well as printers, clothing factories, spice manufacturers, coffee roasters, and metal shops making special parts or doing repairs. Some downtown manufacturers and wholesalers moved to suburbs for lower rents, more space, and easy access to highways. But most plants and warehouses stayed because they provided goods and services to downtown businesses, got their supplies via rail or ship, and wanted to offer an easy commute for workers.[37]

The most familiar piece of this downtown infrastructure was the passenger station, which was usually on the edge of the central business district. Businessmen from out of town exited the station and hailed a cab to a downtown hotel. Local residents picked up friends and relatives who were returning from a trip or coming for a visit. Commuters rode in to work from suburbs. The station was one of the city's most important public spaces because so many people still took trains. The number of riders traveling in parlor and sleeping cars peaked in 1950. Ninety-five percent of adults polled in 1954 had ridden a train, nearly every passenger had taken a smoke amid the inlaid Formica, shining chrome, and curved surfaces of the observation car, and 61 percent of men and 39 percent of women had slept in a Pullman car.[38]

You could get almost anywhere by train. The *California Zephyr* took passengers between San Francisco and Chicago, the *Lark* made trips between San Francisco and Los Angeles, and the Birmingham-Detroit line connected those two cities. The *Lake Shore Limited,* the *Detroiter,* and the *Owl* night train linked cities like Cleveland, Buffalo, Detroit, Providence, and Boston to New York. Kansas City, St. Louis, Louisville, Cincinnati, and Des Moines were all tied to Chicago and one another. A dozen trains still made the trip each day between Manhattan and Chicago.[39]

One of those was the *Twentieth Century Limited,* the pride of the New York Central, which left Grand Central Station every evening at six for its sixteen-hour journey. Passengers boarded by walking along a gray-and-red carpet rolled down a platform known as "the quay." The name of the train line was woven into the rug,[40] and the line itself played a part in Alfred Hitchcock's *North by Northwest*. In one of that movie's many train scenes, Cary Grant says to Eva Marie Saint, "Tell me, what do you do besides lure men to their doom on the *Twentieth Century Limited?*" But with a restaurant, a smoking lounge, and a private compartment, Grant was able to turn his almost certain doom into romance.

As the train pulled into one downtown after another, even Grant had to notice the dingy houses, apartment buildings, and residential hotels that abutted docks and rail yards or stood next to warehouses and small factories. Some residents were poor white laborers who had lived there for decades, others were recent immigrants from the South, and still others were old people without means or motivation to move out. The best-known and perhaps most notorious section of the downtown's residential district was called skid row. Most skid-row residents were white men who, if they worked, washed dishes, served as porters and busboys, and occasionally helped to unload ships. They had less education than the average city resident, tended to be single, and rarely kept in touch with their families. The typical skid-row man either rented a room in a boarding house or cheap hotel, which gave him the luxury of watching television or playing cards in the lobby, or flopped in one of the missions that offered free beds. For entertainment he had nearby bars, a pool hall, a bargain movie house, and a boxing club where he could watch young men spar. For food he had his choice among greasy-spoon restaurants that served hash and onions, spaghetti and meatballs, and bacon and eggs. For clothes, watches, and books he walked to pawn shops, an army-and-navy store, and a place or two selling stolen, fire-damaged, and other distressed goods. He looked for work at a nearby employment agency or went to a designated corner where employers periodically came around in search of part-time laborers.[41]

Like so much else in the '50s, these dilapidated residential zones—and their many broken-down people—remained in check. Many of these men hung around downtown, yet they seemed to know their place, to limit their begging, to restrain themselves. Even in areas "lined almost unbrokenly by cheap hotels and rooming-houses sheltering all manner of strange characters," a guidebook reported, "you won't be in danger." The poor men of New York's Bowery district, recalled William Klein, who took photographs there in the '50s, "didn't shame or threaten the whole city."[42]

The Downtown Style

Downtown in the 1950s was also the cultural center of the city. Women's fashion, popular song, and modern buildings each revealed different aspects of an urban style that was coherent and orderly, romantic yet restrained, luxurious but formal, coming to completion and brimming with confidence. That style was on display downtown.

FASHION

At no other time did the look of women so dominate downtown, and at no other time was that look so coherent. "In the field of fashion," a critic observed of the 1950s, "there is an increasing leveling of the American scene."[43] Such leveling required ready-made clothing, which itself needed a system of regular sizes to fit all female figures. Those sizes were created from a nation-wide measurement of a hundred thousand women in the early '40s. A dominant look also needed a dress code that guided nearly everyone, except for rich mavericks whose tailors and seamstresses made them custom pieces and women whose poverty or personal quirks kept them out of style. The American female, quipped a Parisian, "marches forth with an assurance, independence, and authority that sometimes overwhelm the foreign visitor. But the American woman is not as independent as she appears. The only real competition is for being the first— to look like all the others."[44]

Most women wore some version of the New Look, which first appeared in the late 1940s in the work of designer Christian Dior, whose name, according to surveys, was among the five best-known of the decade.[45] The look featured rounded hips, slim waists, gently curving and sometimes bared shoulders, and full, usually pointy, always protruding breasts. Women like Marilyn Monroe, Jayne Mansfield, Elizabeth Taylor, and Kim Novak set the standard. Like any standard of beauty, the brimming voluptuousness of the '50s was attained by only a few. But most women did dress in ways that took that standard for granted. The average woman, according to a fashion critic, had "taken on a bosom, hips, stomach and derrière. . . . We find ourselves giving up the broad, square, masculine look [of the '40s]. . . . We are slipping into a new form, requiring an hourglass corset." Padded bras, girdles, and zippers (which replaced buttons and hook-and-eye closures after the war) enhanced what *Cosmopolitan* called the look of "rounded femininity."[46] The look was displayed by women on Manhattan's Upper East Side, who followed the fashion scene in *Harper's Bazaar* and *Vogue,* and by working-class girls in Oakland, who took their cues from the advertisements and style sections of *Look, Life,* and the *Saturday Evening Post.*[47]

Form defined the New Look. That form had three components. The first was an emphatic division of top from bottom. This separation was perfected in the evening dress, a nineteenth-century product that went to final extremes in the 1950s. The straps were discarded, the waist was pinched, and the arms, back, shoulders, and armpits were exposed. "Her voice and face, her bosom and hair, her neck and arms are all entrancing," writes a fashion historian, "offering only what is benign among the pleasures afforded by women, all that suggests the unreserved, tender and physically delicious love of mothers even while it seems to promise the rough strife of adult sex."[48] While the top half was on display, skirts covered knees, nylons were standard, and pants were rare. This tension between invitation and rejection was a token of the sexual frustrations of the time. "At night," a woman remembers, "our shoulders were naked, our breasts half-bare, the lower half of our bodies hidden in layers of tulle. Underneath it all, our flesh, like our volatile sexuality, was 'contained' by boned girdles and Merry Widows, in an era when 'containment' was a political as well as a social obsession."[49]

Breasts were the second component of the female form. No other time period displayed such fascination with what a writer called "breast amplitude."[50] Even sketches illustrating short stories in *Reader's Digest* and the *Saturday Evening Post* featured women with tiny waists, big hips, and ample cleavage. The fascination with breasts was encouraged by bras that lifted them higher, separated them further, and made them seem pointy. Advertisements for bras promised "the lift that never lets you down," "lovable curves and superb uplift," and "cream puffs" for girls.[51] The most popular ads were a series for Maidenform. They showed a woman dreaming she wore only a skirt and bra. She was on stage, in a store, or throwing a baseball. In one ad she was placed high atop a pole. From her perch, with firemen and trucks below, she says, "I dreamed I was a fireman in my Maidenform bra. I'm the chief and the siren too—the most incendiary figure in this five-alarm dream! Dangerous, yes . . . but beautifully under control. I'm lifted to new heights of excitement by my dream of a Maidenform."[52]

Incendiary but under control: that was the ideal of female sexuality in the '50s. There was no better example of that ideal than the attitude toward breasts. Fascination with them was nothing new, of course. In the 1930s, for example, Clara Bow showed off their movement and independence beneath silk nightgowns, and the breasts of Maureen O'Sullivan (well, of her body double) moved around naturally as she swam nude underwater in *Tarzan and His Mate*. By the late '40s, however, and increasingly throughout the '50s, breasts were brought into focus more than ever before. In an extreme case of that fetish, movie star Jayne Mansfield prances down the street in *The Girl Can't Help It,* holding two bottles of milk in front of her. As she saunters along, ice melts and eyeglasses shatter. Yet breasts had to be alluring *and* restrained in the '50s. So when an actress of even Mansfield's proportions ran, rode a horse, or descended a staircase in a silk nightgown, her breasts did not move; they came at you, sure and steady, like pointed guns.[53]

The form of the female, lastly, was heavily adorned and layered. Many women wore hats, gloves, and scarves, as well as chokers, bracelets, brooches, and big earrings. Sheer nylons seemed appropriate for the highest heels in history. Slips and girdles were common. Almost 90 percent of women used lipstick, most of them daily: "less than 10 per cent of us," it was estimated, "take a paintless day off each week." That paint was often bright red, though colors did soften late in the decade. Dark eyeliner was turned up at eyes' outer corners, and eyebrows were plucked to a narrowing outer point. All of this created a "smart" or "made-up" image.[54]

Cars have long been emblems of sexuality, but at no other time have the styles of women and automobiles so closely resembled each other. By the early '50s, cars carried more steel and chrome, and displayed more curves and angles, than they ever would again. Ornate grilles and big headlights looked like faces. Some growled, others smiled, a few smirked. Those faces led long, low, sleek bodies that at times were steered symbolically by tail fins. Chrysler called it "the new shape of motion."[55]

The new shape of motion and the look of the ideal female fit together perfectly in the downtown: each was curvy *and* pointy, sleek *and* brimming, stately *and* spirited. Automobile companies made those connections explicit. The Dodge Coronet was advertised as a "Power Packed Beauty," the 1956 Lincoln as a "true sculptured beauty," the Olds 98 as a "sleek new beauty" that combined a "'Go-Ahead' look" with "grace, charm and luxury." Some companies built female features into the car itself. In 1953, for example, General Motors put "bullet-like protuberances" above the front bumpers on several models of the Cadillac. They were named Dagmars, after a well-endowed television actress. *Cosmopolitan* said this about the popular Dagmar: "The movies and Broadway have produced their share of dumb blondes but television is outdoing them with a queen-sized edition who is keeping a lot of men home these nights." Loads of steel and chrome, big engines, bright colors, roomy interiors, a likeness to the ideal female—no wonder the average man had an abiding fetish for his American-made car. "And who can chide him?" asked *Fortune*. "Where and when, as a matter of fact, has the world produced so happy a union of mechanical excellence, sheer elegance, and low cost?"[56]

Americans showed as little interest in the small, homely cars of Europe as they did in the "sack," a loose-fitting dress that cloaked the female figure. According to its designer, Hubert de Givenchy, the sack "is inspired by modern art, the experimental art that seeks new shapes and forms transgressing the limitations set by convention. With my new dress form I have discarded, among other things, the limitations set by the shape of the female figure itself." To discard, in the late 1950s, the limitations set by the female form was an idea ahead of its time. It was ahead of public opinion, too. Eighty-six percent of those polled in 1958 disliked the sack. One man said he had seen "potatoes in better ones."[57]

The sack notwithstanding, women's fashion lent uniformity, a certain formality, even a hint of refinement to downtown. Sure, there were silly hats, rumpled coats, and wrinkled dresses. Plenty of women were overweight; others were clumsy, some unlovely. Yet everyone looked smart. "Back then," a working-class woman told me, "you dressed classy when you went downtown." A college graduate of 1953 later recalled taxicabs "forever unloading magazine editors, who were sometimes ugly but always chic."[58] Just as a person's wealth and education are usually displayed in matters of taste, some cities are more cosmopolitan than others. But the dress code of the '50s prevailed upon nearly every woman who went downtown in an American city. Such a prevalent code, according to a historian of fashion, meant "appropriateness was everything." Leaving the house called for a hat. Going downtown often meant gloves. "Day-into-evening clothes" satisfied the desire "to be properly dressed at all times" and accommodated "the needs of both working girls meeting a date after five and housewives coming into the city for shopping and then the theater."[59]

The formality, the voluptuousness, and the conviction that downtown was worth dressing up for—all lent downtown an air of romance. Downtown attracted many of the city's residents to its numerous bars, restaurants, and movie houses. Night clubs were jammed, and theater was popular. Films, novels, and photographs of the time, as well as today's memories of those years, portray downtown as a central place for wooing. There was a lot of wooing in those years, too, for more of this generation got married, and got married earlier, than any other generation of the twentieth century. Even married people benefited from proximity to the city center. A survey of 730 Detroit wives showed that those with easy access to the "bright lights" of downtown did more satisfying things with their husbands.[60]

Downtown in the '50s, a woman told me, had a "greater sexual spark" than it does today. That seems counterintuitive, for we do not think of formal clothing, reserved manners, and restricted carnal freedoms as conducive to sexual excitement. But, another woman recalled, "Stop-Go lights were flashing everywhere. . . . Sex, its magic spell everywhere, was accompanied by the stern warning: Don't do it!" Alfred Hitchcock once said that Grace Kelly's "apparent frigidity was like a mountain covered with snow. But the mountain was a volcano."[61] Kelly was not your average woman, but Hitchcock's description of her sexuality was true to the times.

There was talk of sex, for example, but it took place in hushed tones or in the cold statistics of Kinsey's report *Sexual Behavior in the Human Female*. Many women had lost their virginity before they got married, but most who did had slept only with their future husbands.[62] Women revealed cleavage, but they rarely wore pants or short dresses. Breasts were more alluring than ever, but they were always holstered. The movies hinted at sex, but they portrayed it only through tempting glances, shared cigarettes, and close-mouthed kisses. The great popular songs of the Gershwins, Cole Porter, Jerome Kern, and Rodgers and Hart also treated sex and romance indirectly. The songs were teasingly oblique. At their sauciest they spoke of someone who was at his best "horizontally speaking," who cherished the "light in your eyes when you surrender," or who confessed, "I'd love to make a tour of you, the eyes, the arms, the mouth of you, the East, West, North, and the South of you." And there was, of course, Marilyn Monroe, who could "suggest one moment that she is the naughtiest little thing," observed Laurence Olivier, "and the next that she's perfectly innocent." Monroe epitomized a general style: "We wore tight, revealing sweaters," a woman recalled, "but they were topped by mincing little Peter Pan collars and perky scarves that seemed to say, 'Who, *me*? Why, I'm just a little girl!'"[63]

Just as clothing reveals what it is supposed to cover, the look of women in the downtown exposed the urban culture of the late '40s and '50s. Most women (and men) obeyed a dress code that made them look more alike than at any other time. They adhered to the idea that downtown was the center of the city and so worth dressing up for. And they behaved with formality and dressed with restraint, even as a voluptuous sexuality lurked on all sides: "Male fashion," according to Anne Hollander, "became more intensely sober, rigid and deliberately reticent, while the feminine fashion business now expanded by promoting and propelling female sexual fantasy back into the vast world of erotic submission and narcissism disguised as modesty, the world of long hair bound up only to be unbound, of tightly girdled waists waiting for male deliverance, of myriad skirts hiding the prizes."[64]

SONG

Irving Berlin, Cole Porter, Jerome Kern, Harold Arlen, Richard Rodgers, Harry Warren, and George Gershwin—working with lyricists like Lorenz Hart, Dorothy Fields, Yip Harburg, Ira Gershwin, and Johnny Mercer—married words to music in songs with the power, as Harburg expressed it, to make you "feel a thought."[65] The thoughts you felt were frivolity, rapture, vulnerability, rejection, loneliness, a crush, the futility of being "just friends." The music was melodic, the lyrics urbane, and together they embodied a simple, but sometimes profound, state of mind. The tune was usually written first, so the lyricist's job was to join syllables to notes, phrases to cadences, and words to musical moods. He had to do so within the constrained form of the thirty-two-bar song and the many abrupt, often ragged musical phrases that, as Ira Gershwin observed, sometimes gave a lyricist little room to "turn around."[66] The perfect fit of words to music, and the feeling of a romantic thought in one of its myriad guises, made these songs popular in their own time and classics today.

These songs arose as a musical form in the 1920s and flourished in the 1930s and 1940s. They reached their peak in popularity, and in style, during the 1950s, when they were given a lush and opulent feeling by being slowed down, set to silkier arrangements, and sung by crooners. The '50s provided the perfect environment for the songs. The check on immigration in the middle of the '20s had weakened ethnic cultures while increasing the population's command of English. The tremendous economic growth and mass market of the late '40s and '50s brought most urban residents into a widening mainstream of cultural life. The quality of recorded music became about as sharp as it is today, and the LP opened up the market for home music. The urban culture of the 1950s, built as it was around downtowns that featured a sartorial decorum, a sensual yet restrained ideal of female beauty, and a heavy emphasis on dating and marriage, responded naturally to the romantic themes, sophisticated melodies, and urbane lyrics of the songs.

"American music," as Irving Berlin called it, drew on a long history of folk songs, marching-band tunes, European opera, vaudeville ditties, religious hymns, blues, and ragtime.[67] That musical history was turned into popular song in the city of the 1920s. "All the old rhythm is gone," said Berlin, "and in its place is heard the hum of an engine, the whirr of wheels, the explosion of exhaust. The leisurely songs that men hummed to the clatter of horse's hoofs do not fit into this new rhythm. . . . The new age demands new music for new action." The new music had to take the quickening pulse of city life, a pulse George Gershwin felt as "staccato, not legato." So the new melodies were stronger and sharper than those of earlier songs, while the wit and verve of the lyrics reflected the sophistication of the urban smart set: art deco, martinis, slick dress, theater, stylish apartment living, new freedoms for women. In a word, the songs represented the rising power of the city over the rest of the country, which eagerly accepted a uniquely urban, mostly Manhattan product at a time when New York City was becoming a national crossroads and every other city's downtown was gaining in stature.[68]

A good song, said Berlin, was "applicable to everyday events." By the 1920s, the nation had become urban enough, and its people connected enough, to have a national music. A wide swath of people heard those tunes in musical theaters and nightclubs, bought the sheet music to play on their pianos, listened to the songs on radio, and, by the early '30s, were watching movie musicals. Berlin referred to that broadening audience, and to the expanding popular culture in cities, when he joked that a good tune "embodies the feeling of the mob."[69]

Berlin also thought that a good song was "easy to sing," "easy to say," and "easy to remember," and therefore it needed to use colloquial American speech. In spite of regional variations and class differences, American English in the '20s and '30s was more uniform across the land, especially in the cities, than virtually any other country's national language. "When it comes to the words they habitually use and the way they use them," H. L. Mencken wrote, "all Americans, even the less tutored, follow pretty much the same line." Lyricists used that common language to reach a large audience. They made special use of the American penchant for creating a lively slang, for making verbs out of nouns, and for ignoring grammatical and syntactical rules and precedent. Ira Gershwin collected urban folklore and listened carefully to casual conversation for slang, phrases, and figures of speech he could use for lyrics.[70] The songs, though witty and urbane, were never stuffy, because they used vernacular expressions and they assumed a core of shared ideas and understandings from everyday life. The combination of wit, restraint, colloquial speech, and common events was personified in the man whom the best songwriters wanted to sing their songs. Fred Astaire epitomized casual elegance. He was formal and proper but came across as a regular guy.

George Gershwin said he wrote songs "of the melting pot, of New York City itself, with its blend of native and immigrant strains. This would allow for many kinds of music, black and white, Eastern and Western, and would call for a style that should achieve out of this diversity, an artistic unity."[71] The songs, in other words, were cosmopolitan: they made you feel a thought that almost everyone else could also feel. The melodies and the words were accessible to anyone who cared to listen. The best of them, in fact, we know today as standards, or classics, that still serve as vehicles for singers and jazz musicians. The melodies and lyrics might well ring as true a hundred years from now as they did seventy years ago. Tony Bennett certainly thinks so: "One thousand years from now," he wrote with hopeful exaggeration, "America will be adored for having created these songs."[72]

Nothing better reveals the universal nature of the songs than the paradoxical fact that most of the songwriters were Jewish immigrants or their sons. Hence that intrinsic American theme of immigrants becoming nationals, of the second generation looking out from their ethnic enclaves and toward the city and the culture at large, of newcomers establishing their place in an open and absorbing culture. These Jewish songwriters did not merely assume the ways of America as their own: they made popular music for the natives of their adopted country.

Singers like Ella Fitzgerald, Tony Bennett, Sarah Vaughan, Nat King Cole, Mel Torme, Dinah Washington, and Perry Como made their reputations from those songs. But the songs made Frank Sinatra more than they made anyone else, because he made them into something no one else could. If formality, romance, and urbanity prevailed in the cities of the '50s, no one personified that mixture better than Sinatra. He did so because he embodied not just its finer qualities, but its darker complexities as well. Periodic tantrums and crude jokes betrayed his formal manners. A contempt for women tainted his grand style of romance. Too often his urbanity was merely a taste for the expensive.

Peter Bogdanovich, who listened to Sinatra in the '50s, remembers his life and his singing as "not only his autobiography but ours as well." Sinatra certainly looked the part: the buffed shoes, the cigarette, the cuffs always perfectly extended just below the sleeves of the jacket. Even his look of studied indifference, which he cultivated in the late '50s—his tie loosened and his jacket hanging by a finger over his shoulder—conveyed a relaxed sophistication. Sinatra wore a hat well into the '60s, saying later on, when you had to justify it, that a man looks right in one "when no one laughs." He was the ideal image of a man in the '50s. He was, of course, a ladies' man who wined and dined his lovers with "class." He was also a man's man. He had brushes with mafiosi. He liked to "duke" waiters with hundred-dollar tips. And he could be crudely shrewd, as Sammy Davis Jr. remembered: "A young cat with two wild-looking chicks walked by and Frank raised his eyebrows. *'Cufflinks.'*"[73] Like so many other adults in the '50s, Sinatra had grown up in an immigrant neighborhood. Even though he moved to Las Vegas and Palm Springs just as working- and middle-class people moved to suburbs, he liked to foster an image of himself as a saloon singer, doing "One for My Baby" at a quarter to three in a downtown bar.

Yet he appealed, like the popular song itself, to nearly everyone. Quincy Jones remembers him in 1957:

All Hollywood was there, and it was really so magical—particularly Frank's entrance into the casino. His name was announced and the crowd began to stand and applaud wildly—Frank was all the way at the back of the room and seemingly not in any hurry to reach the stage. I was leading the band and was afraid the crowd would stop applauding way before he made it to the front. But they didn't. He kept walking in a slow stride, pausing to kiss Grace Kelly and shake Cary Grant's hand. I was so in awe because the crowd got even louder as he paused to take a cigarette, tap it on the cigarette case, and then, in between puffs, he began to sing "Fly Me to the Moon," in between puffs and without missing a beat.[74]

What other singer has been able to enchant the dock worker, the secretary, and the high-class celebrity at the same time? Sinatra's were the top-selling albums from the early '50s to the early '60s, thanks to a mass market in an era that could, for the last time, partake of the old ideals and travails of romance. Angela Lansbury remembers it this way: "In those years, one grew up, fell in love, fell out of love, all to the sound of his voice. A whole generation of people lived with his voice in their lives."[75]

His genius was to take even a corny song, like "Fly Me to the Moon," and render it perfectly for his time. Sincerity, or at least the appearance of sincerity, was a prime virtue in the '50s. Holden Caulfield, in *Catcher in the Rye,* complained about "phonies" and "fakes." Lee Strasberg demanded that performers schooled in the Actors Studio really felt what they portrayed. Sinatra said of himself, "When I sing, I believe, I'm honest. . . . You can be the most artistically perfect performer in the world, but an audience is like a broad—if you're indifferent, Endsville!"[76] The audience believed him when he accented his lines, in "Luck Be a Lady,

Tonight," by blowing in his fist and spilling out dice on cue with the punctuations of the band. You believed because he believed. And he believed as completely in the cynical frivolity of "Just One of Those Things" as he did in the pleasant surprise of "A Foggy Day" and the anguish of "Blues in the Night."

Mel Torme had a richer voice, and Ella Fitzgerald a creamier one, but each sang the words for their pretty sounds. Sinatra sang them for their meaning. More than anyone else, he interpreted a song. He crawled inside its melody and lyric, and he tried to feel its thought with his pace, intonation, and phrasing. "When he sings of love," said a female nightclub performer in 1957, "every woman in the room feels she alone is cuddled in his arms. He doesn't sing only from his throat. Songs bounce off his soul."[77] The silky purr of Sinatra's high baritone acquired an edge in the '50s that let him better express feelings of vulnerability, fragility, and heartache. Sinatra also happened to mature as a man, and as a singer, just when the popular song was ready for its ultimate interpretation in a ripe urban culture. He had lost some hair and put on some weight by the time he turned forty, in 1955. More importantly, he had suffered enough in his short marriage to Ava Gardner to impart an aching beauty to the torch songs. Sinatra's impassioned rendering of those songs was backed by arrangers like Nelson Riddle, who used reeds, strings, a muted trumpet, and a distant-sounding piano to create moods of bliss, solitude, whimsy, relaxation, torment. Before the war, the tunes were often given quirky, light-hearted arrangements and sung in a high, almost operatic voice, with trills and emphatic vowels. They slowed down in the '40s and '50s, acquired more expressive arrangements, and, with Sinatra especially, were sung *to* you, not at you.

The songs, and Sinatra's versions of them, assumed that men opened doors for women, that they tipped their hats when a woman entered the elevator, that the differences between men and women, in other words, were completely natural. So did Gloria de Haven's character in the 1950 movie *Summer Stock:* "A girl doesn't want to be *asked,*" she says to a nerdy guy for whom she has long carried a secret torch, "she wants to be *told.*" Although they sometimes employed clichés about love and the moon and the stars, the songs were also canny and restrained. Lyricist Yip Harburg said, "I doubt that I can ever say 'I love you' head on—it's not the way I think. For me the task is never to say the thing directly, and yet to say it— to think in a curve, so to speak." So it was with other songwriters. Rodgers and Hart were a sweet and sour blend of lush melodies and acerbic lyrics. Ira Gershwin tempered the sensuous energy of his brother's

music with ironic wit. Cole Porter, though gushy at times, was his best
when insouciant and cynical.[78]

The songs dealt with feelings that were true to life and thus emotionally satisfying. They made clever use of word play, lent insight through
metaphor, and thereby enticed the mind. They fused lyric and melody
into a whole that was musically engaging. And as the poetry of modern
urban life, they culminated, along with the economy and culture of the
city itself, in the 1950s.

ARCHITECTURE

While women's fashion and popular songs were vital pieces of a culmi-
nating urban culture that revolved around the nation's downtowns, nothing expressed the 1950s quite as purely as the new office towers. They
were the first tall buildings to rise in over twenty years, and they announced a radical shift in the look and function of downtown.[79] In 1956,
Walker Evans published an article in *Fortune* entitled "'Downtown': A
Last Look Backward." Evans predicted that "the building boom now
commencing will change the face, and a good deal of the atmosphere, of
the whole district." He was right. Just ten years later the typical downtown—which during those ten years had destroyed many of its old
houses, residential hotels, brick warehouses, and small commercial buildings to make way for office towers, parking garages, and convention centers—had what another observer called an "entirely new face."[80]

It was the face of a rejuvenated business sector. In another article a few
years later, *Fortune* described how "a new city embodying the new spirit
emerges with almost incredible rapidity. The accent is on glass—perhaps
because to see, to oversee, to foresee, are headquarters' functions. This
new city is a control tower—tense, watchful, knowledgeable, confi-
dent." The magazine *Architectural Forum* made the same point about a
particular building: "The big, broad-shouldered Chase [Manhattan
Bank] stated crisply the mood and abilities of a newer age. It was not so
much a cathedral of money as a powerful and superbly equipped machine for handling it."[81]

Such images—of downtowns as control towers and of buildings as
money machines—reflected an architectural taste dominated by the themes
of order, precision, and self-confidence. That taste was made possible by
cheaper, standardized, and mass-manufactured building materials. It was
manifest in the typical building of those years, a rectangular steel grid holding sheets of glass—pure lines, in other words. Everything is geometrical

and everything is exact, in some cases right down to constrained blind openings to assure a uniform exterior. Such order came down to one thing: "[T]here has seldom been a period in history," wrote the editors of *Architectural Forum*, "in which problems of pure structure have so prepossessed the architect. Structural logic, structural clarity, structural honesty are almost the controlling values of design." The essence of these buildings is indeed pure structure, even if the logic, clarity, and honesty of that structure are occasionally expressed with tricks like placing glass spandrels between floors to make the entire facade look like glass.[82]

Such reliance on structure, complained Lewis Mumford, ignored the sources that had always inspired building: "[N]ature is one, the cumulative process of history and historic culture are another, and the human psyche is the third."[83] Mumford was right. Whereas older buildings of brick, stone, and terra cotta seem to have come out of the earth, those of steel, glass, and aluminum seem to have been riveted down, right there, from some factory in the sky. The neglect of nature as a source of aesthetic inspiration reflected the conventional wisdom of the time that newer is always better, that there is a technical solution to every social problem, and that the purpose of the environment is to supply resources and receive waste. Americans abused nature on a colossal scale in the '50s: drivers tossed garbage from cars, engineers dammed rivers unnecessarily, scientists tested hydrogen bombs above ground, and the factories of the world's mightiest industrial power spewed untreated waste into rivers, lakes, and the air.

Just as these buildings ignored nature, so too did they neglect the past. So great was the confidence in the power and autonomy of the age that these modern buildings shunned the decoration that had always acknowledged "the cumulative process of history and historic culture." They were built, moreover, on lots that had been cleared with barely a thought for what was standing there. To photographer Walker Evans, the "methodical, almost loving destruction of a building seems to answer a deep human need that is surely akin to humor, to impudence, and to the balm of irreverence. Hence the rapt sidewalk attendance at spectacles of demolition. Who has not cheered the loosening of a dignified old cornice? Or glowed inwardly over the nicely calculated fall of a really solid brick wall?"[84] Evans wrote that in the 1950s, when society gave full play to the innate human capacity to delight in the destruction of a building.

Neither did the buildings of the '50s and early '60s respect what Mumford called the aesthetic needs of the psyche. The surface of these buildings, instead of engrossing the pedestrian's gaze, deflects it, reflects it back, or lets it pass right through the huge panes of glass. If the eye likes to play over carved stone, an elaborate cornice, or wrought-iron filigree, it is disappointed. If the soul wants a building to mesh with the street and its neighboring buildings, rather than stand aloof or tower over them, it is left cold. If the mind desires an impression of harmony given by a pleasing mix of parts, textures, and shapes, again it is frustrated.

The buildings of the '50s and early '60s make different kinds of aesthetic appeals. The exclusive focus on structure appeals to a taste that, above all else, gives priority to the whole. Individual parts like a cornice, or detailed decoration around an entryway, do not really exist in modernist buildings. The materials themselves, mostly glass and steel, adorned at times with bronze or aluminum, do not invite the eye to linger on a particular spot. To appreciate these buildings, you must stand back and see them as pure form. These modern structures also engage the person who appreciates the translation into architecture of leading values of the age, in this case the values of self-sufficiency, precision, and rationality. The translation of those values into cold materials, straight lines, and austere facades makes for a hard, aloof, even arrogant kind of architectural beauty. It is an engineering sort of beauty, stripped of historical references, keen to inspire feelings of lift, power, and sleekness, and heedless of psychological needs for small details, pleasing contrasts, manageable scale, and warmth of material. "The modern architect," quipped a writer in *Architectural Forum*, "did not 'invent' modern architecture. The engineers did."[85]

The buildings that successfully expressed their own brand of modern aesthetics garnered high praise. *Fortune* used words like "spectacular" and "aspiring" to describe the best of them.[86] Even Mumford admired the burnished and bronze-sheathed grid of Mies van der Rohe's Seagram Building and the huge glass walls, gleaming steel, and high ceiling and open floor of the Manufacturers Trust building.[87] Architecture critic Ada Louise Huxtable also admired the very best of the new buildings. While she lamented the demolition of old mansions and apartment buildings along New York's Park Avenue, she saw beauty in their replacements:

As the old buildings disappear radical new ones rise immediately in their place, and the pattern of progress becomes clear: business palaces replace private palaces; soap aristocracy supplants social aristocracy; sleek towers of steel-framed blue, green, or gray tinted glass give the avenue a glamorous and glittering new look. In a surprise shift, elegance has moved from domestic to professional life, from the apartment house to the office building.[88]

Many other writers and most pedestrians, however, could not agree. "For perhaps the first thing to say about the new architectural mode," complained John Updike, "is that it leaves one with little to say. It gloss-ily sheds human comment."[89] For those who work in them, cautioned writer Marya Mannes,

their spanking modernity must be as gratifying as their manifold conveniences. But if their captivity under glass, their insulation in space, does them no harm, I wonder what it does to the outsider who walks by daily. He may be on a strin-gent visual diet that impoverishes his eyes, matched by an emotional malnutrition that contracts his spirit. Whether he knows it or not, he is missing the gentle prods of pleasure that the little old crummy rows, the occasional mansions, the fancy facades once gave him. He has no roughage.[90]

His diet was restricted, thought the editors of *Architectural Forum,* be-cause the "interest in structure has led not to a rich and wide range but to a very narrow range of structural forms." The pedestrian, in other words, saw less variation in the shapes of buildings. He saw less variety of mate-rials, textures, and decorative patterns. And he saw buildings that had to shine like new to flaunt their glassy honesty and steely logic yet showed an unfortunate tendency to get prematurely shabby. While it was true, the architect Edward Durell Stone conceded, that old symbols of wealth, cul-ture, and progress cannot be continually employed, "just as obviously, the world of plate glass and aluminum that is upon us leaves the average mor-tal deeply unsatisfied."[91]

Such dissatisfaction may explain why company presidents and board chairmen filled their new offices with old Sheratons, model ships, and flo-ral prints, and why executive suites in modern buildings like Lever House and the Seagram Building were furnished, not with sleek and austere fur-niture, but with Chippendale chairs and satinwood desks purchased through antique dealers. It may also explain why paintings in the style known as abstract expressionism adorned so many lobbies.[92]

The alliance of modern architecture with abstract expressionism is, at first sight, a strange one. To the untrained eye, abstract paintings are a hodgepodge of swaths, splatters, gobs, and streaks. Even art critic Clement Greenberg, who more than anyone tried to build an audience for these paintings, admitted that "the ultimate source of value or qual-ity" in the canvases of painters like Jackson Pollock, Barnett Newman, Mark Rothko, and Clyfford Still was "not skill, training, or anything else having to do with execution or performance," but rather intuition or in-spiration. Perhaps that is why so many paintings were unplanned: "[W]hen the painting is finished," said painter William Baziotes, "the subject reveals itself." For Robert Motherwell, too, painting was "an adventure, without preconceived ideas." A modern building, to the contrary, does not suggest disorder, unplanned adventure, or a lack of technical skill. But it may have needed some roughage. So, at least, thought *Fortune* magazine: "Businessmen willing to settle for color (often globs of it), texture (ranging all the way from paste-ups to stitched burlap), movement, and light have found that abstract paintings can, in fact, provide a sense of emotional release, and may give the beholder a thin grip on humanity in a business-machine world." Huge abstract canvases, in other words, "make the functional buildings warmer and more inviting." Possibly. Or maybe owners and tenants just wanted the latest thing in art.[93]

Then again, perhaps there really was more there than met the eye. Just as modern architecture of the 1950s turned its back on history and nature, so too did abstract paintings. "The image we produce," said Barnett Newman, "can be understood by anyone who will look at it without the nostalgic glasses of history."[94] He was asking that the modern painting, like the modern building, be seen just as it is, out of context, without reference. Similarly, just as modern facades replaced brick and stone with glass and steel and swapped earth tones for tinted blues and greens, so did abstract paintings replace concrete images of things and people with abstruse speckles, dollops, and swirls. There was yet another correspondence between the canvases and the buildings. Many modern paintings, including some of the most famous ones, were "'all-over,'" in Clement Greenberg's words, "that is, filled from edge to edge with evenly spaced motifs that repeated themselves uniformly like the elements in a wallpaper pattern." Many modern buildings similarly displayed repeating geometric patterns that ended simply because the building ended. The resemblances between these buildings and paintings emerge in the following passage from Greenberg, in which I've placed in brackets the architectural terms corresponding to those of abstract painting: "Not only does the abstract picture [modern building] seem to offer a narrower, more physical and less imaginative kind of experience than the illusionist picture [decorated building], but it appears to do without the nouns and transitive verbs, as it were, of the language of painting [architecture]."[95]

It takes work to appreciate both modern buildings and abstract expressionism. You do not recognize things or people among the stripes and swirls. Nor do you see a link to history or to nature in the material or design of a glinting skyscraper. To appreciate an abstract canvas or a glass-and-steel building, you must engage it intellectually—which means learning how to talk about it. You are not asked to like it; you are meant to understand it. To understand it, you must make a case for it.[96]

A case must be made for such modern art forms because they have taken abstraction to an extreme. Just as architects stripped down their buildings to give an impression of sheer lift, order, or power, the abstract painters eschewed images for patterns, streaks, and globs meant to incite a feeling or strike a nerve. Such extreme abstraction neglected classic ways of appealing to the senses and asked the mind to assert interpretation, detect meaning, and search for feeling. Critic Leo Steinberg wondered whether he was imposing an analytical scheme on the paintings of Pollock and Jasper Johns, and whether in trying to impose meaning he was saying more about himself than about the art: "It is a kind of self-analysis that a new image can throw you into and for which I am grateful. I am left in a state of anxious uncertainty by the painting, about painting, about myself. And I suspect that this is all right."[97] For the critic of abstract painting or modern architecture to feel all right about his state of anxious uncertainty is one thing. To expect the ordinary city dweller to feel good about it is quite another.

That the modern building was an exercise in abstraction seems fitting. Downtown, after all, was about to become more abstracted from the life of its city. That the modern building was equally an exercise in aesthetic control seems ironic. For the central business district was about to lose its pull on the neighborhoods, its grip on the suburbs, and its power to make the city whole and to express the city's style.

City Hall and the Transformation of Downtown

Portland, Oregon, Mayor Earl Riley called the postwar years "a new era in municipal administration." The "new" meant an end to the old. America's cities, reported *Life* in 1955, were "full of worn-out old political machines, many already tossed on the scrap heap, others wheezing along in the last stages of obsolescence." New Deal programs like Social Security and unemployment compensation had been slowly replacing the welfare functions of political parties. Civil service had reduced the number of jobs controlled by the party in office. The foreign immigrants who fueled party machines were in short supply after the war. A growing middle class was moving to suburbs, taking charge of state legislatures, and imposing constraints on city governments. Ed Flynn, chairman of the Bronx Democratic Party for thirty years, described the mood in the late '40s:

"[T]housands of our young people are being taught, in high school and college . . . that machines themselves are unnecessary to begin with and remain only as vicious anachronisms, that the 'good' citizen must never become a part of, or even support, such wicked monsters."[98]

Whereas the winning party had customarily treated city hall as its very own job bank, favor dispenser, and reelection headquarters, the urban regimes of the late '40s and '50s broadened into coalitions of businessmen, civic leaders, urban planners, civil servants, party hacks, and reformist mayors. In Pittsburgh, for example, wealthy Republican industrialists and financiers cooperated for the first time with Democratic city politicians. Chicago's Democratic Mayor Richard J. Daley enjoyed the support of leading Republican businessmen and the four Republican-owned newspapers. Mayor Richard Lee came up through the New Haven Democratic Party, but he was neither its servant nor its master. He courted the city's business community, delivered enough patronage to keep his troops in line, and procured huge sums of urban-renewal money from the federal government.[99]

Such alliances formed throughout the 1950s in cities all across the country. Businessmen wanted a share in building highways, convention centers, and downtown office towers. More educated and professional people took jobs as park commissioners, traffic engineers, and redevelopment experts. Groups of concerned citizens wanted city hall cleaned up and downtown spruced up. The parties lost some of their control over election slates and became vulnerable to independent and even amateur candidates. Less crooked and more efficient men occupied the mayor's office: Chicago's Daley, New Haven's Lee, Detroit's Albert Cobo, St. Louis's Raymond Tucker, Milwaukee's Frank Zeidler, Pittsburgh's David Lawrence, New Orleans's de Lesseps S. Morrison, and Philadelphia's Joe Clark and Richardson Dilworth. Some of these men were political novices, and others had risen to power through the machines. They all ran city governments increasingly open to outsiders, less prone to graft, and freer from party control. Cities in the '50s all over the United States, in other words, junked most of the old machine and reformed what was left of it.[100]

The decline of the machine, wrote *Life* in 1955, spelled the end of "the oldtime political boss." Not all bosses were bad. The Bronx's Ed Flynn, for example, who died in 1953, was regarded by *Life* (and in other accounts) as "an exceptional boss—a man of education and personal charm, loved by countless friends and respected by opponents."[101] But even Flynn's version of his own activities shows why many people wanted to scrap the machines. As boss of his party's executive committee, Flynn had his district leaders "taken care of" by ensuring that they, and sometimes their family and friends, got jobs "exempt" from civil-service requirements. He helped those with civil-service jobs gain promotions. He ignored neighborhood opposition to lucrative projects like an inner-city highway. Patronage was the nature of politics for men like Flynn, and they got it and dispensed it through their political machines.[102]

Few bosses were as benevolent as Flynn. Thomas McCoy ran Providence between 1936 and 1945; his brother ran it for the next five years. Widespread voting fraud, open gambling, and the mayors' cozy ties to one of the two city papers eventually created enough resentment to bring down the brothers' regime. In 1937, Frank Hague declared himself "the law in Jersey City." He remained the city's "overlord" until 1948. E. H. Crump ran Memphis for over twenty years. Crump decided who could open a business, who got city patronage, and who served as sheriffs and judges. The beginning of his end came in 1948, when Governor Estes Kefauver helped citizen groups in Memphis write a civil-service law, establish a permanent register of voters, place voting machines at polling stations, and use black election officials in mostly black wards.[103]

City hall in the '50s was something between the old-time machine and today's more bureaucratic style of administration. Parties relinquished control, bosses lost clout, and outsiders filled top positions, including the mayor's office. In many cities, the mayor and his party still cut back-room deals and delivered votes and patronage, though on a smaller scale than twenty or thirty years earlier.

New York City illustrates a common scenario: a machine that had no real boss and was losing power fast. William O'Dwyer's term as mayor, which began right after the war, was the last gasp of the old world. O'Dwyer liked to denounce the Democratic Party machine, but he was of it, worked with it, and depended on it. His own administration, in fact, was a modern version of old Tammany Hall. Graft was widespread; O'Dwyer appointees served time for extortion; policemen and politicians profited from waterfront crime; and O'Dwyer himself retired in 1950 when information surfaced about his ties to Frank Costello, Thomas Luchese, and other underworld racketeers.[104] Vincent Impelliteri took over as interim mayor. He was a longtime ward heeler who became city-council president thanks to his talent as a party yes-man. When Impelliteri ran for mayor later that year, he was backed by one clique of the party against another. His faction played to the political mood of the times, and on the O'Dwyer scandal as well, by running him successfully as an "anti-boss," "anti-politician" candidate.[105]

Carmine De Sapio took over New York's Democratic Party when it was losing its power to nominate candidates and amass patronage, and went on to make it more efficient and get the applause of city papers and national magazines. "Where the old Tammany used to pass around food baskets," observed *Time*, "De Sapio's Tammany makes public-minded donations to blood banks. Where the old bosses packed city hall with hoodlums and hacks, De Sapio helps to find good men."[106] There were still hangers-on and "riff-raff" in a party not yet weaned of graft, but even *Fortune,* the business magazine, thought New York was well run in the '50s. Talented men took office, the City Planning Commission acted vigorously, and the city's police, fire, and health departments got high marks. Yet the party machine was so creaky by 1961 that Robert Wagner, who was denied the nomination of the Democratic Party, ran in opposition to it and won.[107]

In Philadelphia, as in many other cities, a political outsider upset the ruling party and established a cleaner and more efficient government. Philadelphia had one of the country's few Republican machines in the '30s and '40s. But like the Democratic regimes elsewhere, it padded payrolls, took kickbacks from local manufacturers for reduced water bills, pressured city employees to give money for mayoral campaigns, and allowed ward bosses to organize numbers and slot-machine rackets.[108] Many agreed with the *Saturday Evening Post*'s depiction of Philadelphia as "the most lackadaisical city in America, and one of the crookedest." Then, in 1952, Joe Clark was elected mayor. Clark had no political experience, and he was not a party man. The city of Philadelphia, wrote the *Post,* had "at last blasted free from the grip of the old-fashioned bosses and is trying, with frantic zeal, to reform itself."[109]

Clark was a war veteran, and veterans were key figures in the postwar years. They were characters in movies, the government rewarded them with generous benefits, and the public was asked to be patient with them as they returned to civilian life. Veterans were also thought to embody such virtues as honesty, duty, and responsibility, and some were ready to carry those virtues into politics. "It took six post-war years," according to the *Evening and Sunday Bulletin,* "for Joe Clark and his fellow-veterans— such men as Richardson Dilworth of the Marines, James A. Finnegan of the Army Air Force, and Robert K. (Buck) Sawyer of the Army Engineers—to wrest the city government from a Republican regime that had held it for more than half a century." These new men did a lot of good. During Clark's tenure, from 1952 to 1955, city services improved, graft declined, successful prosecutors jailed corrupt men, civil-service rules governed the city's twenty-five thousand jobs, and five black lawyers joined the DA's office.[110]

Entrenched interests checked other reforms. James Finnegan was a big man in the party and president of the city council. Although he jammed a reform charter down the throats of unwilling Democratic ward leaders, he also had to procure and distribute enough patronage to keep his troops in line and his party in office. To that end, Finnegan fought successfully against such reforms as eliminating residence requirements for city jobs, which would have allowed Mayor Clark to hire outside people for his government. Even in Philadelphia, where the ruling party had been ousted and an amateur politician was in charge, patronage was still necessary to run the city and stay in power.[111]

Chicago's was a reformed but still powerful machine in the '50s. Between 1933 and 1947, Mayor Ed Kelly ran the country's mightiest party organization. During those years, the party controlled candidate slates and paid off politicians and judges, the board of education was indifferent, the streets were dirty, and the police were even dirtier.[112] The next mayor, Martin Kennelly, expanded civil service and relinquished some patronage. But he enacted few other reforms, and aldermen, policemen, and ward bosses had long leashes: "[T]he graft of the ward committeemen," admitted an alderman in 1951, "it's brutal."[113] Kennelly looks like an honest, if naive, man when compared to the bosses of what *Harper's* called the "graft-laden regimes that preceded him." But like so many other administrations during the late '40s and early '50s, Kennelly's was seen as "a transition stage between government by machine and government by law."[114]

Richard J. Daley may not have turned Chicago into a city of law, but he did curb its worst kinds of corruption and was, by all accounts, personally honest, even as he looked the other way from much that went on around him. Daley called himself "the first of the new bosses." He disciplined his machine by tightening the reins on local ward heelers and precinct captains. He satisfied a growing middle class with better police, and improved city finances and the handling of major construction projects. He gained the public's confidence, in other words, by carrying out his political maxim: "[G]ood government is good politics and good politics is good government."[115]

Yet *Time* called Daley the "boss-mayor." He had come up through the party ranks and did many things the old way, like picking candidate slates, distributing patronage, and commanding nearly all fifty aldermen on the city council. His party controlled some ten thousand payroll jobs. It influenced the awarding of contracts to build office towers, urban freeways, and public housing. It allowed money to change hands when the lack of a building permit was ignored, or traffic tickets were "lost," or votes were bought on election day.[116] It tolerated illegal and secret connections between the mafia, high-ranking police, and city-hall insiders.[117] It even controlled Chicago's black neighborhoods, which were organized by a local party apparatus headed by a black politician who was tied to the big white machine downtown. So it was for good reason that old Paddy Bauler, one of Chicago's last saloon-keeper politicians, called Daley "the dog with the big nuts."[118]

In the late 1950s Chicago, New York, and Philadelphia, as well as cities like Providence, Jersey City, Cleveland, Pittsburgh, Kansas City, Memphis, New Orleans, St. Louis, and New Haven, still had at least some working parts of once more powerful machines. Other cities had only remnants of their old regimes. By the second half of the '50s, for example, Detroit's mayor had lost most of the patronage available to his predecessors. Boston's James Michael Curley, a former mayor, governor, and big-time boss, no longer had the stuff to reclaim the mayor's office in 1955. In Newark, a *set* of power structures, rather than a single power structure, ran the city by the end of the decade.[119]

Some cities had never had bosses, ward heelers, and patronage machines. These were usually cities whose most important bursts of population growth, city building, and industrial development came after World War II, and that therefore never had the immigrants and the manufacturing sectors that combined between the 1870s and 1920s to create the classic urban machines. Cities like Norfolk, Charlotte, Atlanta, Dallas, San Antonio, Denver, Albuquerque, Tucson, Phoenix, Los Angeles, and San Jose were examples of what *Fortune* called a "businessman's city." Businessmen liked such places because they were run by growth-oriented administrations that improved the efficiency of city government and built new highways, airports, office towers, subdivisions, and water and sewer systems. The men governing these mostly southern and southwestern cities were the same men found at meetings of the Jaycees and in the chamber of commerce board room. They rarely dealt with precinct captains or labor leaders.[120]

The modernization of city government throughout the nation in the 1950s meant a new relationship with voters. Whereas precinct captains used to deal directly with neighborhoods, more residents now went downtown to deal with bureaucrats. Civil-service exams decided who got the jobs that were formerly acquired in neighborhood bars. If the awarding of some construction projects still required kickbacks from local businessmen, many more contracts were awarded on an open and competitive basis. Mayoral candidate Richardson Dilworth whipped up spontaneous street-corner rallies during his failed 1948 challenge to Philadelphia's Republican machine. By the middle of the '50s, according to Edward Banfield, "candidates for mayor must catch the eye of the public as best they can. Each sponsors several large rallies and all buy a good deal of television and radio time and newspaper space, plaster the city with billboards, bumper stickers, posters, and brochures, and fill its air with sound-truck oratory."[121]

James Michael Curley personified the changes. Curley ended his long career in Boston politics in 1955. His tactics were out of date in a time when campaigns were starting to be waged as much on television as in the streets. Yet he walked the wards, attended the wakes and funerals of ordinary citizens, visited the docks to address a fishermen's union, heard complaints and requests of voters in his house, got city jobs for friends of friends, and tried to shore up support among precinct bosses in the old Irish, Italian, and Jewish neighborhoods. Curley's personal style failed to get him elected in 1955, but over the years it connected him to many of Boston's citizens. In 1958, tens of thousands of mourners filed by his bier, at a rate, it was reported, of four thousand an hour.[122]

As city hall moved from a world of bosses and machines to one of experts and televised campaigns, it also played an important role in starting, during the 1950s, what eventually became a radical transformation of downtown. The central business district needed new office buildings. Clogged streets called for parking garages, one-way traffic flows, and inner-city highways that would slice through or curve around the edge of downtown. The move of some manufacturers to suburbs, and the use of trucks to haul freight once carried by ship and rail, emptied out a few downtown warehouses and factory buildings. Old produce markets on the edge of the central business district were busy in the early morning but nearly empty during the day. Planners considered them an inefficient use of urgently needed space. The sense of urgency was intensified by the recognition that downtown stores, restaurants, and nightclubs were losing customers to shopping centers, just as downtown theaters and movie houses were losing patrons to television.

To preserve downtown as an attractive place that unified the city was a daunting task, but decision makers and planners in those days believed strongly in progress. Nearly every mayor, journalist, and city planner agreed with the chairman of the nation's Central Business District Council: "[T]here is no problem in any city," he said, "that cannot be solved if citizens of good will join together to solve it." Encouraged by federal programs (enacted in 1949 and 1954) to provide legal means and heaps of money for urban renewal, almost every city had a planning commission and a redevelopment agency by the middle of the '50s. Planners studied downtown traffic flows, port facilities, and commercial and industrial potential. They proposed to demolish warehouses, residential hotels, and commercial buildings that they deemed decrepit, just plain old, or simply in the way. All sorts of schemes depicted new residential high-rises, convention centers, and modern office towers for the downtown. Men in crewcuts, white shirts, and horn-rimmed glasses explained it all with charts and graphs and models. New was better, they thought, just as uniformity was superior and straight lines and open spaces were more rational. And it all seemed so efficient.[123] City planning, writes an urban historian, "has never been more effective than when it was harnessed to the goals of that powerful coalition of progressive business elements and activist chief executives that took shape in city after city following the Second World War and that exploited the potent planning tools made available by the federal urban renewal and highway programs to lay claim to the decaying urban core."[124]

Lay claim they did. St. Louis began ripping chunks out of its city center during the '50s to make way for new parks, parking garages, and high-rises. Developers razed entire blocks of downtown Philadelphia to create what the *Saturday Evening Post* called "new beauty spots, highways and business projects." New Haven tried to stimulate downtown shopping by building a highway into its central business district. Downtown Pittsburgh was full of old warehouses, its streets were clogged with cars, and there was no room to build. The water of its converging rivers stank and the soot of belching steel mills blackened its buildings. Migrants from the South moved into decrepit tenements on the edge of downtown. So Pittsburgh, like St. Louis, Philadelphia, New Haven, and almost every other big city during the '50s, scrubbed its buildings and started to clear space for parks, parking lots, new apartment buildings, elevated highways, office towers, and convention centers.[125]

Such renewal gave us the downtowns of today. A highway slices through a part of almost every downtown in America. Empty lots are the sole reminders of buildings demolished forty years ago. Central business districts that once afforded all manner of commerce, services, and entertainment have become finance centers catering to the daytime needs of office workers. Most downtown sidewalks are empty on Friday nights and Saturday afternoons. Decorated buildings of brick, stone, and terra cotta, so pleasing to the pedestrian, are fewer in number and are overshadowed by office towers that look their best when seen at night from a speeding car.

Those trends have played out differently across the nation. Some downtowns, like those of Gary, Newark, and Kansas City, are deserted on weekends, full of empty lots, and suffused with melancholy. Cities like Baltimore, Boston, and Denver, on the other hand, have used new libraries, ball parks, and shopping plazas, along with the conversion of warehouses into stores and condominiums, to lift their downtowns out of the doldrums of the 1970s and 1980s. A few downtowns, like Chicago's and San Francisco's, cater to daytime workers, draw shoppers and visitors on weekends and house large numbers of residents in new apartment buildings and converted factories and warehouses. But even San Francisco's lively downtown "is nothing, absolutely nothing," a delicatessen owner told me, "compared to when I opened in 1952."

It is hard to imagine how downtowns could have gone another way in the 1950s and 1960s. Explosive growth in the corporate and white-collar sectors demanded new buildings for central business districts that had not erected a skyscraper in twenty years. Cities had little choice but to expand their downtown office space if they wanted headquarters, construction jobs, more daytime office workers, and greater revenue from business taxes. Nor did they have much choice but to erect tall buildings of glass and steel. After the war it became easier and cheaper to build higher. Fine decoration in brick and terra cotta was too costly. And the spirit of the time—the faith in technology, the neglect of nature, the intense desire for rational control, the belief that the present had no need for the past—had to be inscribed in the new buildings.

Every big city, wrote an urban planner in 1951, also had to find ways "to bolster the downtown business district so that it will not decline and become ineffective in competition with outlying shopping centers." Bolstering downtown's retail sector meant, in large part, making it easier for cars to get there. The number of riders on public transport peaked in 1947. Then the car took over. The economy was booming, gas was cheap, and automobile companies (and the United Auto Workers) were at their peak, while teamsters and trucking companies were on the rise. The typical young couple, meanwhile, wanted a car, two if possible, in the driveway of their single-family house in the suburbs. The federal government secured mortgages on those new suburban homes and also provided big money to state governments, which spent the funds on freeway construction, and to local governments, which spent the money on acquiring land for downtown parking garages.[126]

Added to all of this was a dogmatic faith in whatever was new, modern,

and efficient, a faith that justified the clearance of "slums" from the edges
of downtown. Such areas were full of old houses, residential hotels, and
apartment buildings. They were usually inhabited by the men of skid row,
by black and white migrants from the South, and by longtime residents
who could not or did not want to leave. Some buildings were beyond repair, but many just needed paint jobs and new electrical and plumbing
systems.[127] To rehabilitate shabby housing inhabited by poor people
would have required creative financial incentives by government. It was
easier to clear entire blocks for on-ramps, highways, parking garages, convention centers, new apartment buildings, or, as happened in many cities,
lots that would remain vacant for decades. Few people appreciated the
beauty of older structures on sometimes crooked and cramped streets, or
the virtues of mixed land uses and building types.

By the late 1950s, there was no stopping the increase in automobiles,
the spread of suburbs, the rise of new office towers, the decline of passenger trains and street cars, and, therefore, the decline and fall of downtown as the center of metropolitan commerce, politics, and entertainment. Redevelopment agencies were loaded with federal money,
equipped with the power of eminent domain, and staffed by planners
who saw apartment houses, poor people, old office buildings, and commercial side streets as barriers to progress. Labor unions, construction
companies, and trucking firms saw wages and profits in skyscrapers and
convention centers. Many young couples left city neighborhoods for suburbs and visited shopping centers instead of downtown stores. No one at
that time wanted to live in converted warehouses on the edge of the central business district. Few cities opposed the building of urban freeways
by state and federal agencies. Even in San Francisco, where there was
organized resistance, an elevated freeway rimmed the downtown's waterfront by the end of the '50s. What Stefan Lorent wrote of Pittsburgh
applied to nearly all American cities: "The town has no worship of land-
marks. Instead, it takes pleasure in the swing of the headache ball and the
crash of falling brick."[128] Even Manhattan's Pennsylvania Station got the
ball. After the majestic building was pounded into rubble, a *New York
Times* editorial lamented an age that "will probably be judged not by the
monuments we build but by those we have destroyed."[129]

Despite the changes on the horizon, however, the downtown of 1955
still looked more like the downtown of 1925, or perhaps even that of 1905,
than it did the downtown of 1965. In 1955, modern buildings stood out
against a shorter, warmer, decorated architectural background of brick,
stone, and terra cotta. Ten years later, the modern style had begun to

obliterate varied textures, rich ornamentation, and myriad nooks and crannies. Downtown in 1955 was still an intricate and stylish world of retail stores, business offices, and places of entertainment. Ten years later it was losing its department stores to shopping centers, its movie theaters to television sets, its nightclubs to rock and roll, its small commercial buildings to urban renewal, and its urbanity to informality. In the middle of the '50s, the edges of central business districts merged with apartment buildings, residential hotels, and corner stores, eateries, and laundries. By the middle of the '60s, many of those nearby residents and merchants had been driven out to make way for convention centers, new high-rises, parking lots, and vacant lots. In 1955, freeways were "becoming the framework for a new city" and were starting to break up the "intimacy and compactness" of downtown.[130] Ten years later, they *were* the framework of a new city, and they *had* fractured the intimacy and compactness of downtown. Whereas the rail station was still a point of civic pride and common experience in 1955, it was nearly forgotten a decade later. And even though plans for overhauling downtowns were drawn up during the mid-'50s, most of those plans were put into practice only five or ten years later.[131]

There was some truth to the prediction, made by a planning magazine in 1950, that changes then underway "will not be on a scale large enough to make the city of 1960 look very different from the city of 1950."[132] Yet it was also true that downtown was changing, bit by bit, throughout the decade: a row of brick buildings came down, a modern tower went up; a parking garage opened, a dozen streetcars retired; a hat shop moved out, a pawn shop moved in; a movie theater shut down as fewer people went downtown. Those individual events chipped away at downtown and fractured the unity of the city. By decade's end they had created enough of a pattern for Jane Jacobs to sound a prescient warning: "Without a strong and *inclusive* central heart, a city tends to become a collection of interests isolated from one another. It falters at producing something greater, socially, culturally and economically, than the sum of its parts."[133]

FIGURE I. A street corner in Manhattan, late 1940s. Photograph by Arnold Eagle. Courtesy of the Museum of the City of New York.

FIGURE 2. A young couple on their lunch break in San Francisco. The newscopy depicted them as typical of white-collar workers after the war. Courtesy of San Francisco History Center, San Francisco Public Library.

FIGURE 3. Shoppers in downtown San Francisco, 1956. Courtesy of San Francisco History Center, San Francisco Public Library.

FIGURE 4. A street in downtown Worcester, Massachusetts, early 1950s. Photograph by George Cocaine. Courtesy of the Worcester Historical Museum, Worcester, Massachusetts.

FIGURE 5. The heart of Oakland's business district, early 1950s. Courtesy of the Oakland Public Library, History Room.

FIGURE 6. New York's Times Square, 1953. Photograph by Andreas Feininger. Courtesy of the Museum of the City of New York.

FIGURE 7. A sleek Buick in Oakland's business district, mid-1950s. Courtesy of the Oakland Public Library, History Room.

FIGURE 8. A man delivering flowers in downtown Manhattan, 1955. Photograph by Frank Paulin. Courtesy of the Museum of the City of New York.

FIGURE 9. A glass-and-steel tower in downtown Oakland, late 1950s. Courtesy of the Oakland Public Library, History Room.

FIGURE 10. White-collar workers in front of Pittsburgh's new U.S. Steel–Mellon Building, 1953. Photograph by Clyde Hare. Courtesy of the Carnegie Library of Pittsburgh.

FIGURE II. A secretary in a modern office building at Pittsburgh's Gateway Center, 1953. Photograph by Harold Corsini. Courtesy of the Carnegie Library of Pittsburgh.

FIGURE 12. The caption on this March 31, 1959, photograph published in a San Francisco newspaper read: "These diners at a restaurant—Tiki Bob's—here are enjoying cheesecake provided for them by the management, but it isn't the edible kind. Restaurant owner Bob Bryant has run up his lunches from 45 to over 450 daily by providing 'fashion shows,' with heavy emphasis on lingerie. Police issued an edict to have the shows stopped, claiming 'there is a time and place for everything.' But there was no raid 3/30 as model Rene Mills displayed a swim suit." Courtesy of San Francisco History Center, San Francisco Public Library.

FIGURE 13. Commuters getting off trains in San Francisco, 1949. They were, according to the newscopy, "hustling across the Third and Townsend-sts tracks to catch Muni streetcars and buses for their offices." Courtesy of San Francisco History Center, San Francisco Public Library.

FIGURE 14. Waiting room at New York's Grand Central Station, 1952. Photograph by Larry Silver. Courtesy of the Museum of the City of New York.

FIGURE 15. Piers on Manhattan's West Side, ca. 1950. Photograph by Andreas Feininger. Courtesy of the Museum of the City of New York.

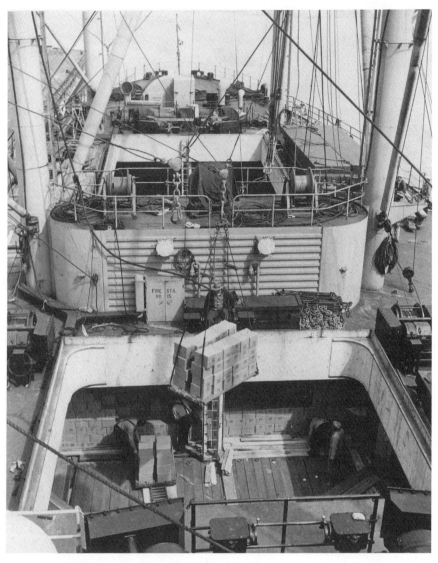

FIGURE 16. Workers unloading a ship in San Francisco's port, late 1950s. Courtesy of San Francisco History Center, San Francisco Public Library.

FIGURE 17. Piers, cranes, and rail cars in Weehawken, New Jersey, ca. 1950. Photograph by Andreas Feininger. Courtesy of the Museum of the City of New York.

FIGURE 18. Skid row on the south side of downtown San Francisco, 1953. The newscopy read: "CLEANUP DUE?—Here's a part of Skid Road. Supervisors last week asked Dr. Ellis Sox, city health director, to start a survey of health and sanitation law violations, and proposed also that study be given to a cleanup-paintup campaign for the rundown district." Courtesy of San Francisco History Center, San Francisco Public Library.

FIGURE 19. A new parking garage in downtown Pittsburgh, 1953. Photograph by Harold Corsini. Courtesy of the Carnegie Library of Pittsburgh.

FIGURE 20. The tip of Pittsburgh's central business district—called the Point—was full of rail yards, warehouses, and small factories in 1950, before redevelopment a few years later. Courtesy of the Carnegie Library of Pittsburgh.

FIGURE 21. The original caption for this 1954 photograph read: "A popular symbol of Pittsburgh's progress: Through ruins at the Point, the camera focuses on the stainless steel of Equitable Life's Gateway Center, the city's first redevelopment project." Photograph by James Blair. Courtesy of the Carnegie Library of Pittsburgh.

FIGURE 22. The Point, in 1959, with its new Gateway Center, Hilton Hotel, and park. Pittsburgh was the nation's most radical example of redevelopment in a central business district. Courtesy of the Carnegie Library of Pittsburgh.

CHAPTER 3

The Neighborhoods

"So complete is each neighborhood," observed E. B. White, "and so strong the sense of neighborhood, that many a New Yorker spends a lifetime within the confines of an area smaller than a country village." Writer Marya Mannes detected a "structure" to the neighborhoods: "basically triangular" and consisting of "the relation of side street to avenue, of residence to commerce, of privacy to common experience." To live in that structure, she felt, "is to be part of an intimate complex of people and services that form, as time goes on, a close familiar whole in the midst of the great fragmentation which is the city." A recent memoir shows the old neighborhood lending "a constancy, a consistency, a dependable texture to everyday life."[1]

Such images pervade memoirs about the 1950s, as well as books and magazines written at the time. Neighborhood stores, after all, were small and were owned by those who worked in them. Mothers stayed at home with their children. Sisters and uncles lived a block over or a few streets away. Households were connected to local schools, churches, union halls, and retail streets. One's neighborhood, in other words, was basically the same thing as one's community. It was typically a prosperous community, too, for the wages of both blue- and white-collar workers rose steadily between the late '40s and the early '60s.

If the average neighborhood was in its prime during the '50s, it was also at a turning point. Small stores faced competition from the first supermarkets. Manufacturers started to abandon the city. Young couples moved to suburbs. Adolescents began to define themselves in opposition to adult taste and, in some cases, to adult authority. For the most part, however, these were piecemeal and gradual changes that coalesced into an obvious pattern of decline in white neighborhoods only during the later '60s or the '70s. Many neighborhoods that housed black people, conversely, faced a rapid and devastating transition in the early '60s. People there lost a lot of blue-collar jobs. Middle- and upper-class residents took advantage of opportunities to move into other parts of the city and the suburbs. Entire blocks were demolished to make way for public housing—or to remain vacant for decades. All of that turned poor but vigorous black neighborhoods into desperate ghettoes by the mid-'60s.

Yet as long as the old neighborhoods lasted, and they lasted right through the '50s, the themes of continuity and coherence prevailed over those of movement and fragmentation. These themes prevailed in the entrenched white neighborhoods, where this chapter begins, and in the growing black communities, too, where it ends.

Neighborhood Commerce

Small stores, most of them run by their owners, drew people from the neighborhood's residential streets toward its main avenue, thereby creating a commercial focus for the area. Those stores supplied just about all of the area's everyday needs, created a stable and personal world for shoppers and merchants, and made local retail streets, or clusters of corner stores, the centers of neighborhood social life. The commercial scene alone made the typical neighborhood of the '50s a very different world from today's.

Most big cities had at least one residential district made up of tall apartment buildings. Corner stores dominated every intersection. In some places, nearly every block was lined with shops, stores, restaurants, and bars. These were "virtually self-sufficient" commercial worlds, writes E. B. White:

A man starts for work in the morning and before he has gone two hundred yards he has completed half a dozen missions: bought a paper, left a pair of shoes to be soled, picked up a pack of cigarettes, ordered a bottle of whiskey to be dispatched in the opposite direction against his home-coming, written a message to the unseen forces of the wood cellar [where you placed your order for ice, coal, and wood on a pad outside], and notified the dry cleaner that a pair of trousers awaits call. Homeward bound eight hours later, he buys a bunch of pussy willows, a Mazda bulb, a drink, a shine—all between the corner where he steps off the bus and his apartment.[2]

A more common type of neighborhood was made up of three-, four-, and five-story apartment buildings. Residents walked no more than five or six blocks to a major shopping street. In the Bronx, for example, East Tremont Avenue ran right through the center of the Tremont section of the borough. The avenue was "a bright, bustling mile" of butchers, bakers, delicatessens, laundries, coffee shops, and drug, grocery, clothing, and hardware stores.[3] Until the 1950s, Brooklyn's Pitkin Avenue had "crowds of shoppers and strollers, day and evening." With its "banks, Woolworth's, classy shops, loan companies, Loew's Pitkin, the Yiddish theater, the Little Oriental restaurant," wrote Alfred Kazin in 1951, "it might be Main Street in any moderately large town." A man remembered Pitkin Avenue, after it had become the run-down commercial street of a slum, as "the best shopping street you could go to in the whole world. It was be-eee-you-ti-ful!"[4] On Tremont and Pitkin, and on thousands of other commercial streets just like them, you bought your daily groceries, made an occasional purchase of furniture or clothing, met friends for coffee or lunch, went to the drug or hardware store, watched a movie, or had a drink.

The retail street gave focus to lower-density districts as well, like those comprised of row houses, bungalows, Victorian two-flats, or detached "triple-deckers." San Francisco's Richmond district was full of abutting houses and small apartment buildings on street corners. "Clement is the street," observed a journalist, "that unites them as a community. It is a street so warm, so *gemutlich,* so pleasant to shop, that first-time visitors often feel they have made a private discovery."[5] The same could have been said of the commercial street that united St. Nick's parish in Chicago, a neighborhood of small, closely spaced, brick bungalows. The life of St. Nick's centered on Sixty-third Street, a retail strip that was especially dense and lively toward either end of the neighborhood. The street had a Walgreen's and a Kresge's, which were chain stores, but also a bank, a movie theater, a men's store, a shoe store, a candy store, a sporting-goods store, and a grocery store, all independently owned by local businessmen. Shoppers walked to these commercial clusters. Women wheeled shopping carts in the mornings, kids made late afternoon runs for fresh dinner bread, and couples and entire families took strolls on Monday and Thursday evenings, the weekday shopping nights. Such streets were open for at least one and often two nights a week in almost every big city during the '50s. Staying open at night was a way for neighborhood stores to compete with downtown nightclubs, department stores, and movie theaters, as well as with new shopping centers in suburbs.[6]

Some of the neighborhoods made up of closely spaced single-family houses did not have a retail street and were served, instead, by an "oasis of local shops" on certain street corners. "Our neighborhood life," recalls Doris Kearns Goodwin, "converged on a cluster of stores at the corner of our residential area: the drugstore and butcher shop; the soda shop, which sold papers, magazines, and comics; the delicatessen; and the combination barber shop and beauty parlor." Bert Kemp's memoir *EB* describes a retail corner in Brooklyn that had, on one side, McDade's Saloon, Joe's Barber Shop, and Pease's candy store; across the street was Dubin's Drug Store, the Ideal Meat Market, Vick's Grocery, Wenn's Deli, and Gray's Real Estate: "Booze, beer, companionship, cigarettes, egg creams, ham sandwiches, pain relievers, and a shave and a haircut . . . what more could you ask? A man could live his entire life on that corner. And some men damn near did."[7]

Not all neighborhood shopping districts were prosperous, pleasant, or safe. Plenty of poor whites lived near docks and rail lines, among warehouses and small factories, and in tenement districts that had housed several waves of immigrants. Such places had their share of decrepit buildings and littered sidewalks. Along the retail streets, writes novelist Irving Shulman, "blowzy housewives in soiled Hoover aprons and pajamas, over which they wore old coats frayed at the cuffs and hems, plodded to the bakeries, butcher shops, vegetable stands and groceries." Many of those old coats came from stores that sold "second-hand clothes, faded blouses and torn sweaters, uncalled-for laundry, yarns, threads, buttons, candles, penny hardware items, and old hats, coats, and ties." Some merchants allowed criminals the use of their bars, lunch counters, and pool rooms to run numbers or launder money. But mafiosi tended to be discreet, one man remembered, "and we weren't too aware of them."[8] If hoodlums generally kept to themselves, they didn't think twice about roughing up a bookie who wasn't paying off, as in this scene from Shulman's novel *Cry Tough!*:

The proprietor of the candy store, a balding man in a soiled apron, was placing packs of cigarettes in the racks on the wall behind the counter when the three of them entered. At the end of the counter, alone, hunched over the racing page of the next morning's *Daily Mirror,* sat a man who needed a shave. He wore a stained gabardine suit, with soiled jacket elbows, a blue sport shirt open at the throat, a light woven hat pushed back on his head, cheap white rayon hose with brown clocks, and brightly polished slip-on moccasins. The bookie's lips were blue in the neon light and his sharp nose appeared to be ferreting information out of the paper. . . . Larry's right fist, clenched around the roll of pennies, piled into the back of the bookie's head.[9]

Despite differences in neighborhood size, wealth, or density, the commercial scenes in most neighborhoods shared certain characteristics. One common trait that lasted throughout the '50s was the European flavor of many shops. Small bakeries featured the specialties of the neighborhood's dominant ethnic group: bagels, baklava, strudels, ryes, or rolls. Bars served the beer and hard liquor of the old country. Signs in German, Scandinavian, Polish, or Italian announced the restaurant's lunch specials. Plate-glass windows were lettered with the proprietor's name and shaded by a big canvas awning that said "delicatessen." Jewish delis had no ham or shrimp salad, of course. Instead, they grilled frankfurters and knishes in the window and filled their glass cases with rows of smoked meats, trays of garnishes, and bottles of soda and beer.[10]

The commercial life of neighborhood stores was more direct and tactile than it would be in later years. The butcher cut your meat and the deli prepared its own dishes. The grocer sliced, measured, and weighed your order. He then took a thick-leaded pencil from behind his ear to

tally the prices on the side of a brown paper bag plucked from the stack under the counter—he almost always knew the right size for the "order." Once the addition was done (and rechecked by adding *up* the column) he would lift his fingers like a concert pianist's and press down a group of keys, ringing up the total on his ornate, manually operated, cast-iron-and-bronze National Cash Register. . . . *Gzing*-GZING went the register, up into its glass-faced top went the numerical total . . . and out popped the wood-slatted cash drawer.[11]

The butcher, deli owner, and grocer, as well as the hardware man and candy-store proprietor, knew more about their products than most store employees do today. So did the neighborhood's shoe-repair man, tailor or seamstress, and fixer of watches, clocks, radios, or television sets. Many proprietors hung a neon sign to announce the owner's name and the store's function—Jack's Men's Shop, Hank's Jewelry and Watch Repair, Goursau Meats, Brownie's Hardware, MacMinnis Stationery, Ratner Paints, Frank's Shoe Repair, Dino's Delicatessen, Ictor's Coffee Shop. The store owner, his daughter, or an artist who lived in the neighborhood painted window signs that advertised daily or weekly specials. Storekeepers naturally arranged their goods in some fashion, but without the extreme uniformity typical of today's chain stores. Many shoppers, according to a book about New York City, liked "to handle the profusion of merchandise" that was often "swinging overhead or stacked on every available inch of space."[12]

Because so many stores were run by their owners, relations between customers and shopkeepers often became what a New Yorker at the time described as "friendship as well as habit." That is also how Doris Kearns Goodwin remembered it forty years later. The druggist, deli owner, and grocery man, she wrote, were "as much a part of my daily life as the families who lived on my street."[13] Nearly everyone remembers the neighborhood stores as integral parts of their lives. Take Pease's candy store, described in Kemp's memoir *EB*. On your left as you entered Pease's was the cigar and cigarette counter; on your right were the comic book and magazine stands. Farther in on the left was the soda fountain, with its swivel-topped chairs, marble counter, and steel spouts. On the right side ran the penny and five-cent candy counters. Then came the back tables. The owner was Malcolm Pease. He was, according to one of the patrons,

a benevolent dictator. He was a nice man who truly liked and enjoyed the kids, but he, like anyone else, had his limits. He would clear the place out when on a whim (or more likely a realization that his store was very crowded but he was not making very much money). AWRIGHT! EVERYBODY OUT. . . . Fifteen or twenty kids would dutifully spill out of Pease's candy store onto the sidewalk and congregate in small groups, talking and waiting. . . . Pretty soon the place was just as crowded as before.[14]

Barbers, butchers, and grocers, like the owners of candy stores, drugstores, and delicatessens, watched the children grow up.

If merchants smiled at much of what they saw, they also shook their heads at some of it. Customers got on their nerves. A few teenagers shoplifted. A husband liked to insult his wife at the checkout counter. The heavy drinkers wanted more booze on credit. Plenty of shoppers, by the same token, disliked particular store owners. Some of them fell into foul moods, now and again yelling at a kid flipping through a comic book: "Hey, ya gonna read or ya gonna buy?" Others poked fun at certain customers, sometimes cruelly. A few were too burdened by life to care. Mrs. Sanew, for example, worked behind her candy-store counter seven days a week, from seven in the morning until ten at night. Writer Pete Hamill remembers her "pinched sour mouth" that knew no laughter. Every neighborhood also had its "empty-handed barber, the clerk in the antique store nobody ever comes into, the idle insurance salesman, the failing haberdasher—all of those," observed John Cheever, "who stand at the windows of the city and watch the afternoon go down."[15]

In spite of a failing business, some rude merchants, and a few shoppers who shunned routine chit-chat, the commercial scene in the neighborhoods of the 1950s was overwhelmingly one of small shops run by owners who valued personal relationships with customers and of shoppers who valued their familiarity with merchants. Most women, like this one in Seattle, stuck to their local retail streets:

I like to shop here in our neighborhood. . . . I try to buy at little places I know—where they will make things good for you and they won't be snotty. I may pay more sometimes by not looking for bargains, but the stores I know are good and if anything goes wrong they will take it back. I get personal attention and service because they know me. For example, I buy meat from a butcher I know—the meat isn't any cheaper there, but he always gives me choice cuts. Whenever I'm buying furniture I go to this one store in the neighborhood. I got my chest of drawers there a while back because the salesfellow was so darn nice.

Working-class women were especially tied to their local merchants. As a whole, they were less likely than middle- and upper-income women to classify themselves as "regular shoppers" in the downtown or as frequenters of new shopping centers in suburbs. A working-class woman who did like to shop downtown avoided "any of those small type shops" where the salesladies "try to make you feel that you can't walk out without buying something. They make you feel awful if you say you don't like something they have told you is nice. And they would certainly think it was terrible if you told them that you didn't have enough money to buy something."[16]

The "salesfellow" who was "so darn nice," mentioned in the quote above, may have been unctuous. He may have turned to his partner, after a certain shopper left his store, and said, "Man, what a pain in the ass she is." But he, like she, lived in a commercial world where shopkeepers sold on credit, steady customers got better goods, and buyer and seller developed some trust. They talked about the merchandise, the weather, the new baby. The talk was superficial, but it was a predictable and genuine kind of superficiality. Even banks wanted close relations with customers. On Friday evenings or Saturday mornings, for example, neighborhood kids went to the local bank to fill their savings accounts with whatever remained from allowances, paper routes, and baby-sitting. Adults in one Chicago neighborhood went to the bank on every twelfth Friday to have quarterly dividends stamped in their books. There was no need to do this—dividends were automatically credited—but the bank encouraged the ritual to tighten its links with the community.[17] Many shoppers valued familiarity with merchants because it gave them a feeling of trust, turned shopping into a minor social occasion, and made some of them think they were getting a good deal. Merchants, as they still do today, valued friendly relations with customers because they were good for business.

Itinerant peddlers, such as knife sharpeners and old-clothes buyers, added another dimension to this personal world of neighborhood commerce. They made weekly or monthly rounds along the side streets of many working-class districts. The clothes buyer, according to a magazine piece, yelled out, "I buy! Cash clothes!" He worked neighborhoods where the "housewives know him. When they hear his familiar cry, and they have clothes to sell, they lean from apartment windows and call, 'Hey, mister!'" A few customers, one buyer said, "look down on you. They show they don't like to deal with a peddler." Some bargained relentlessly and sent him away if they couldn't get their price. Others settled on a price easily and offered a chair and a cold drink on a hot day. The old-clothes buyer had to be friendly with the local tailor, druggist, or shoe-repair man, with whom he left his heavy package of clothes when he got to a neighborhood. After making the rounds, he returned to collect his stuff and took the trolley downtown, where he sold his goods to second-hand clothes dealers.[18]

The local bar was as much of a neighborhood fixture as the deli, laundry, barber shop, produce stand, hardware store, funeral parlor, and pharmacy. The bartender knew most of the customers, bought a few of them a drink now and then, and went occasionally to a ball game with a couple of regulars. The porter was often a neighborhood hanger-on who cleaned up after the bar closed. If there was food, it was served by a local woman who cooked spaghetti and meatballs and maybe some roast chicken. Many bars participated in community activities. A Chicago cab driver, who in 1963 ended his career as a "neighborhood tavern keeper," talked about his former bar not only as a watering hole, but as a supporter of Little League baseball and a sponsor of a local softball team: "Every Sunday we'd have a game with a team from another neighborhood. We'd play for a half-barrel of beer. Then, after the game, they'd come over to my tavern and drink it."[19] Other bars participated in less benign activities. The heyday of the saloon-keeper politician was long gone by the 1950s, but many tavern owners still had enough clout to obtain illegal building permits, fix parking tickets, get a Christmas turkey for a poor family, and procure jobs for a few of the locals as revenue clerks or park workers. Bar owners did such favors through "petty dictators" who were known to "run the show" in many working-class precincts. The recipient of a favor paid for it with his own vote and with those of his friends and relatives. Some neighborhood bars participated in even shadier exchanges, like peddling stolen property or engaging in payoff schemes with local cops, teamsters, and the association of jukebox operators.[20]

Most neighborhood bars served men only. Some men stopped in each day for a couple of beers on their way home from work. A few spent Friday nights swapping war stories and watching the local fights on television. Others spent Saturday and Sunday afternoons watching baseball teams whose starting lineups, like the patrons of the bar, varied little from one year to the next. Adolescents became men in neighborhood bars by drinking with their buddies or by taking a girl to the one bar on the main street that had a little "class," thanks to its quiet back room or private booths. Every bar had its problem drinkers who routinely made their wives and kids nervous when they were late arriving home from work. On paydays, especially, many a young son was sent to the corner bar to fetch his dad. One of those boys later recalled how "it embarrassed his father to see his little son edge his way into that smelly saloon, hesitant, tentative, a little frightened, but determined to stick it out. And it didn't hurt any that the other men urged him to go home with his son."[21] The scene in the bar was not always pleasant, but like nearly every other aspect of neighborhood commerce, it was always personal.

Stoops, Mothers, and Teens

Writer Pete Hamill remembers trips downtown, to the beach, and to the other side of the city. "But it was to the Neighborhood that we always returned. Other neighborhoods were not simply strange; they were probably unknowable. I was like everybody else. In the Neighborhood I always knew where I was; it provided my center of gravity." The neighborhood provided residents their center of gravity because it was like a city in miniature. "In every neighborhood," remembered Chicago's Mike Royko,

could be found all the ingredients of the small town: the local tavern, the funeral parlor, the bakery, the vegetable store, the butcher shop, the drugstore, the neighborhood drunk, the neighborhood trollop, the neighborhood idiot, the neighborhood war hero, the neighborhood police station, the neighborhood team, the neighborhood sports star, the ball field, the barber shop, the pool hall, the clubs, and the main street. . . . So, for a variety of reasons, ranging from convenience to fear to economics, people stayed in their own neighborhoods, loving it, enjoying the closeness, the friendliness, the familiarity, and trying to save enough money to move out.[22]

The neighborhood was a bigger place back then because it offered almost everything residents needed and so included more of their lives than it does today. It was a smaller place, too, because it was so enclosed, even isolated, and therefore somewhat provincial and confining. Retail streets, full of owners who ran their own small businesses, certainly tied residents to their neighborhoods and helped create a sense of belonging and self-sufficiency. But it took more than commerce to consolidate the tightly knit communities of the late '40s, '50s, and early '60s. It also required (among other things) mothers at home, residents keeping company on stoops and porches, and teenagers at ease in an adult world.

A lot of social life took place on stoops. In a Chicago neighborhood, for example, a dozen people would sit in front of a house conducting several conversations at once or just watching the street. "You could have gone down the alley," a resident remembered of summertime, "and walked in the back door of every house and robbed people blind. Everybody was out on the front porch." More than 80 percent of the houses in another Chicago neighborhood had benches or couches on their porches in the late 1950s. So it was in other cities. "Everyone had a well-used stoop," a New Yorker remembers, where people talked, drank, bad-mouthed certain neighbors, watched the small children, encouraged stickball batters, and teased teenagers as they left for dates. Former residents of Philadelphia and Baltimore recall row-house porches full of women and kids during parts of the day. In San Francisco, neighbors chatted on the steep wooden steps of Victorians, while along Brooklyn's Eastern Parkway "men and women sat on benches and collapsible chairs, and caught up with the gossip."[23]

Most residents sat outside to watch their kids play, see who walked by, or just wind down with the fading light. Some took to steps and porches because their houses and apartments were cramped or stuffy. Others did so because their dwellings were closed to outsiders: it is impossible, wrote an observer of a working-class neighborhood,

to conceive of [many homes] as something other than a "woman's world." The curtains are of lace, the bedspreads are chenille, the furniture elaborate with design and doilies, the colors are light and delicate. Here, there is simply no place for a male with his dirty hands, informality, and coarse manners. . . . Ultimately, the home is a kind of ceremonial center used only at those rare moments when visitors are permitted to look into family life.

For all those reasons, neighborhood residents made good use of their stoops, porches, and sidewalks until at least the middle of the '50s. By then, most households had a television set, and "on summer nights," remembers

Pete Hamill, "the streets were emptier, as each apartment lit up w.. pale blue glow."[24]

If many houses and apartments were a "woman's world," so too, in many ways, was the entire neighborhood. The vast majority of women did not earn wages. During the day, most women were either at home, shopping on the nearby retail strip, or visiting somewhere on the block. A woman describes life in her middle-class apartment building:

The fathers went to their paid jobs early each morning, while we mothers organized days of housekeeping, child care, and plans for the evening, when the breadwinners would return. Mothers met in the lobby in the morning to take kids to the park, ending up eating lunch together. Once or twice a week a play group met in someone's apartment, mothers taking their turns overseeing the kids one week in order to free a couple of afternoons the next week.[25]

The stoop, the block, the park, the retail street, the apartment building— these were the public places of women's lives in the neighborhood. A typical working-class mother, for example, met her longtime girlfriend on Thursdays at the local coffee shop and visited regularly with neighbors, a nearby in-law, or her sister or parents who lived a few streets away. One in four working-class women belonged to a local association of some sort. Mothers with school-age children tended to join the PTA. Some wives belonged to groups related to their husbands' work, like "fire ladies" or "policemen's pals." Many others lent time to their churches and synagogues. They went to weekly services, of course, and they (and some of their husbands) joined the church bowling team, volunteered to chaperone dances for teenagers, and helped organize summer trips for the children. In Catholic neighborhoods, especially, the parish—which offered weekly masses, parochial school, community-service groups, and links with the city's political machine—was nearly the same thing as the community. Like working class women, middle-class women also met friends on the retail strip, volunteered at their church or school, and visited with neighbors and relatives. But they were more likely than women in the working class to go out of the neighborhood for a class or a reading group, volunteer at a museum, join the Rotary-Anns or Ki-wanitias, or meet friends downtown for lunch and shopping.[26]

Mothers at home meant order in the household. The evening meal, for example, was a critical part of family life. The man came home from work, the kids assembled, the food was served. The meal meant continuity, a family routine, control over the children. It did not always mean happiness. Plenty of tyrannical fathers cowed their children, just as countless meals were eaten in anxious silence, broken only by the scraping of knives and forks and the occasional command to sit up straight or chew before swallowing. Whether pleasant, boring, or painful, the evening meal was rarely missed. Nor did the average family dine out often. There were fewer neighborhood restaurants back then, and fast-food franchises had yet to change the way Americans eat.

Mothers imposed order on public life as well. They constituted, along with the local merchants, the neighborhood's informal police. Youngsters knew that the neighborhood mothers, as well as the merchants who knew those mothers, watched almost everything they did on the block, at corner stores, and along the retail street. There was plenty to watch. High birth rates after the war resulted in a lot of kids in the late '40s and '50s. Those children participated in fewer organized activities than they do today, and so had more friends in the neighborhood and made regular use of stoops, alleys, sidewalks, and side streets as play areas. A kid's world, it was said, had two parts: the "block," where every stoop and alley had been explored and every neighbor and shopkeeper were known, and the alluring yet unknown "beyond," which was everywhere and everyone else.[27]

Like their younger siblings, adolescents knew every nook and cranny of the neighborhood. They also knew their places in it. Some boys fought, others bothered girls, and a few had foul mouths. But by today's standards, teenagers followed adult rules, manners, and tastes. They grew up in a time when a parent, instead of negotiating or calling a "time-out," could point a finger and reply, "Do what I say." Teachers, too, had much more authority than they have today.[28]

Yet if teens were still underlings in an adult world, their status began to shift in the '50s, and it shifted most rapidly in the neighborhoods. The small town did not usually have enough teenagers to generate the critical mass for an adolescent culture, and most suburbs were full of younger kids. Cities, on the other hand, had the necessary number of teens (those born in the '30s and early '40s), the high schools and retail streets where they could gather, and the theaters and ballrooms where they could dance and—in the late '50s—see rock-and-roll acts.

The incipient separation of youth from the larger culture astonished adults. "We have always had pubescents and adolescents," wrote Dwight Macdonald in a 1958 article on youth culture. "But now we also have something quite different; namely, teenagers—not just children growing into adults but a sharply differentiated part of the population." While older generations had gone through their own phases of adolescence, this new generation was seen by adults as acting out their problems. Teens also began looking to themselves, rather than to adults, for standards of taste and behavior. James Coleman observed in 1961 that "the old 'levers' by which children are motivated—approval or disapproval of parents and teachers—are less efficient." Even those who saw no decline in ethical standards did notice a fall in "the *manners* of many teen-agers."[29] A look at those manners—the clothes teens wore, the places they gathered, and the music they listened to—reveals a teenage world that, while not yet a thing unto itself, started to alter the character of neighborhoods, the nature of family life, and the structure of American culture in the 1950s.

Adolescent manners changed because, for the first time, most teenagers finished high school and thus spent four years in an incubator of teen culture. Teenagers also had money in their pockets, thanks to allowances, after-school jobs, and baby-sitting. That money generated a new and growing market for adolescent clothing, makeup, and magazines. *Seventeen,* for example, which had a circulation of almost five million in the late '50s, ran advertisements on over 90 percent of its pages. *Mad* magazine sold a million copies, and a lot of advertising space, by the middle of the decade. Teenagers bought sixty million comic books each month during the '50s. Television studios and advertisers saw adolescents as a large and impressionable audience.[30] And, perhaps most importantly, teenagers were fast becoming the single biggest market for music.

Their relationship to music was like their relationship to society: firmly tied to adult standards. Until the middle of the '50s, for example, almost all teens who listened to popular music listened to the same music as their parents. The radio show *Lucky Strike Hit Parade,* which started in 1935, played the old popular songs coast to coast on Saturday nights until its demise in 1959. Most of the nation's five thousand AM radio stations, and its five hundred thousand jukeboxes, played those same songs during the early and middle '50s. Even Dick Clark's *American Bandstand,* which did not begin its national television broadcasts until the end of the decade, played juvenile variations of torch songs, country tunes, and blanched blues numbers sung by affable crooners like Guy Mitchell, Frankie Lane, Pat Boone, and Johnny Ray. The kids who danced on Clark's show were small and innocent versions of their parents. The girls were forbidden to wear slacks, low-necked gowns, or tight sweaters on the program. So they dressed in knee-length skirts and blouses. Boys wore jackets and ties. All were restrained and proper.[31]

If most teens still listened to their parents' music, or juvenile variations of it, they did so in their own ways. They listened to jukeboxes at neighborhood diners, soda bars, pool rooms, and candy stores. They went to Friday- and Saturday-night dances in church buildings, high-school gyms, and Knights of Columbus halls, where the first hundred kids sometimes got free hot dogs and Pepsi. Groups like the Catholic Youth Council typically sponsored such dances, which one participant later remembered as "a conservative, community-sanctioned institu-tion" for working- and middle-class youth; there were few black kids or "tougher elements of the working class."[32]

Hi-Teen clubs sponsored dances that were often broadcast live on Sat-urday afternoons from a downtown ballroom or a neighborhood theater. In Buffalo, one or two thousand of the city's twenty-two thousand Hi-Teen members typically attended dances at the Delwood Ballroom. These kids dressed in slightly altered versions of their parents' clothing. Hi-Teen boys wore oversized jackets, dark shirts, and baggy cuffed pants; the girls wore long skirts, pullovers, white socks, and saddle shoes. These pro-grams encouraged kids to keep straight, and they often held special dances to raise money for such groups as the March of Dimes. The per-formers on the Hi-Teen bandstand also played at Buffalo's Town Casino, a popular Main Street nightclub for adults. The kids, like their parents, listened to Tony Martin, Vic Damone, Benny Goodman, and Lionel Hampton.[33]

Some teenagers, of course, listened to an entirely new sound. By the early '50s, nearly every big city had a flamboyant white disk jockey—Dewey Philips in Memphis, "John R." Richbourg in Nashville, Zenas Sears in Atlanta, Hunter Hancock in Los Angeles, George "Hound Dog" Lorenz in Buffalo—who played a lot of black music for mostly white kids. Lorenz broadcast six nights a week from a club where 20 percent of the audience was black. He played Fats Domino, Bill Haley, Little Richard, Lloyd Price, LaVerne Baker, Jerry Lee Lewis, Carl Perkins, Chuck Berry, and Screamin' Jay Hawkins. Disk jockey Alan Freed hosted dance con-certs in Cleveland and New York. Most of the performers were black—Ruth Brown, Joe Turner, the Drifters, Fats Domino—but the audience was nearly half white. Thanks to these disk jockeys, "Crazy Man Crazy," by Bill Haley and the Comets, made the best-selling list on *Billboard*'s na-tional chart in 1953. Two years later, Haley's "Rock Around the Clock"—which Frank Zappa remembered as "the Teen-Age National Anthem"—was number one and sold six million records. Elvis recorded his biggest hits in the late '50s. Between 1955 and 1959, 147 of the 342 top-ten hits were rock and roll. Before it splintered into various genres and markets in the mid-'60s, rock was a fairly coherent collection of songs, bands, and styles that bound together musically a large and diverse audience of teenagers.[34]

Most teenagers who listened to Little Richard, Elvis Presley, and Bill Haley, and who dressed and talked in ways that were slightly different from their parents, still acknowledged the adult rules and hierarchies of the neighborhood. Teenage culture was more about style than values. By today's standards, adolescence in the '50s was a tame period of experimentation before settling down to an early marriage, child rearing, and a house.

It is also true, however, as Hubert Selby describes in *Last Exit to Brooklyn*, that a few local toughs hung out on street corners in lower- and working-class neighborhoods: "A warm clear night and they walked in small circles, dragging the right foot slowly in the hip Cocksakie shuffle, cigarettes hanging from mouths, collars of sport-shirts turned up in the back, down and rolled in the front . . . spitting after every other word, aiming for a crack in the sidewalk." Some teenagers joined youth gangs. They hung around pool rooms, barber shops, and candy stores, and they rented a place now and then for a party with liquor, a band, and girls. Members of certain social and athletic clubs stitched their names on sateen jackets, roughhoused in candy stores and cafeterias, and occasionally robbed a store or held someone up. Almost every local tough had to "defend" the neighborhood at least once or twice from outsiders who came looking for trouble because of a girl or an ethnic or racial slur.[35] Movies like *Rebel Without a Cause, The Wild Ones, The Blackboard Jungle,* and *West Side Story* depicted the teenage rebelliousness that was emerging in the '50s.

What memoirs, magazine articles, and novels of the era all suggest, however, is that even the rougher boys wanted the respect of their parents and were deferential to neighbors and local merchants. In Buffalo, for example, working-class gangs made up of second-generation Polish, Irish, and Hungarian kids engaged in skirmishes that rarely caused injury. A woman remembers one gang, the Gunners, as "really nice guys. My mother and father would sit on the porch in front of our house and the Gunners would walk down the street with a lead pipe on the way to a 'rumble,' and they'd always stop and say hello to my mother and father." That is not to make light of youth violence. But there was little of it in the '50s, and it was child's play by today's standards. Even so, many adults saw the gang fights, roughhousing, dances, and jukeboxes as part of a scourge of juvenile delinquency because, for the first time, teenagers were starting to separate themselves culturally from adults.[36]

Yet most teenagers were optimistic and conformed happily to a conservative mainstream youth culture. For every young person who resisted the authority of his or her parents, and who cultivated a new and separate identity around music and clothes, a hundred others fit comfortably into the world of their parents. No other modern generation, after all, came of age in such a conservative time. Many teenagers had seen an uncle, a father, or a brother go off to fight in Europe, the Pacific, or Korea. They had watched Senator McCarthy bullying innocent people on TV. They were taught to "duck and cover" from bombs. Teenage girls, according to the research director of *Seventeen* magazine, "tend to date earlier, marry younger, accept more responsibility than their mothers did in their own teen years. Sure, they follow fads, but the main interests of the teenage girl today are her family, her home, her education, her career, marriage, and a home of her own."[37]

Settled Down

Americans have always been mobile, and the '40s and '50s were no exception. Tens of thousands of people moved to cities in search of jobs during the war. By the late '40s, over a million families a year were making the great trek to suburbs. Migrants from the South jammed into existing black communities in northern cities. As black districts ran out of living space, their residents expanded into adjacent white areas. Thus began the familiar pattern of a black family moving onto the block, the posting of "For Sale" signs, and the inevitable "white flight." Some neighborhoods took in other kinds of newcomers. Boyle Heights in Los Angeles, for example, accommodated Mexicans. Manhattan and Newark received Puerto Ricans. Parts of the near North Side in Chicago, the Village in Manhattan, and the Haight-Ashbury and North Beach neighborhoods of San Francisco housed college students, homosexuals, and aspiring artists.[38]

Despite such movements to, within, and out of cities, white neighborhoods during the '40s and '50s were rather stable when compared to other times. One man recalled that people in his neighborhood "stayed put; nobody moved."[39] Plenty of people moved, of course, but there is some truth in what he remembered. Few foreigners (relative to earlier or later times) entered American cities during the '40s and '50s. Expanding black districts did not directly affect most white neighborhoods, though whites everywhere did perceive the expansion as a threat. The loss of city residents to suburbs had little immediate effect on the neighborhoods. Most of those leaving were young couples who had been either renting apartments or doubling up with parents or siblings. Their loss would be felt only ten or fifteen years later, when some neighborhoods found themselves with a lot of retired people, fewer adults in the prime of their working lives, and fewer children. Retail streets had yet to suffer the drop in demand that would come with the loss of manufacturing jobs in the '60s. Nor did they yet face serious competition from the suburban shopping centers and large chain stores that would later drive out local pharmacists, grocers, bankers, and butchers.

Continuity, then, was a dominant theme in most white neighborhoods during the late '40s and '50s. In Pittsburgh's Italian and Polish districts, for example, about a quarter of all houses in 1960 had a resident who had been living there for at least thirty years. Many houses without such long-standing inhabitants were occupied by the former owner's son or daughter, who was usually a young parent. We tend to think that every young couple moved to suburbs during the '50s. But in Pittsburgh, at least, the number taking over their parents' city houses exceeded those moving to suburbs. And of those who left their childhood houses but stayed in the city, nearly half continued to live in the neighborhood. "The pull of the ethnic neighborhood and the services it provided," concluded the authors of a study about Pittsburgh, "obviously remained strong. Even the so-called attractions of the suburbs and the threat of a black incursion that supposedly caused many to flee the city in the 1950s failed to lure the second-generation immigrant from his neighborhood."[40]

That analysis agrees with casual comments, like this one, made in magazines at the time: "The neighborhood, in North Philadelphia, is lower middle-class economically, but most of the families own their own homes. Most of the men work at skilled trades and most of the women stay home and take care of their houses. Many of them grew up in the neighborhood." Even those who moved to suburbs kept ties, for a while at least, to the old neighborhood. Young couples visited their parents or siblings on weekends. Churches and synagogues were crowded with former residents during holidays. Some suburbanites still shopped on the old retail street for ethnic foods or because they knew the shopkeepers. Local clubs had dues-paying members who no longer lived in the neighborhood of their youth.[41]

Most neighborhoods were quite settled down in the '50s. Merchants and customers got to know and sometimes trust each other. If most neighbors were not friends, they had grown accustomed to one another over the years. Bonds were strengthened by parents' embellishments on the neighborhood tales they passed to their children. A grocer could playfully remind a young mother, who struggled with her willful toddler in his store, what a little brat she had been twenty years earlier. Most of the neighborhoods had settled down ethnically as well. That is not to say cities lost their ethnic flavor. Many, in fact, were still accretions of little ethnic nations. Detroit, for example, was described in a 1956 magazine article as "a city of pockets" with "a variety of nationalities and races." The "old identifications" and the "pressure of old prejudices" still bound people to their neighborhoods. "Go that way, past the viaduct," Mike Royko wrote of white Chicago in the 1950s, "and the wops will jump you, or chase you into Jew town. Go the other way, beyond the park, and the Polacks would stomp on you. Cross those streetcar tracks, and the Micks will shower you with Irish confetti from the brickyards. And who can tell what the niggers might do?"[42]

But ethnic identities and territorial loyalties slackened after the war. "No one can say that all is love and kisses in this grand mixture," wrote Jacques Barzun in 1954:

there are two sides of the railroad track and on one side the poorer group, very likely ethnic in character, is discriminated against. But at what a rate these distinctions disappear! In Europe a thousand years of war, pogroms and massacres settle nothing. Here two generations of common schooling, intermarriage, ward politics, and labor unions create social peace.

The grand mixture in American cities had always meant the amalgamation and eventual blanching of incoming strains of European cultures. But that mix, as Barzun implies, had become a nearly uniform blend by the 1950s. Only five million foreigners entered the United States between 1926 and 1960. Five million came between 1850 and 1870, another five million came in the 1880s alone, and one million came each year between 1906 and 1915. The percentage of foreign-born residents in a typical city was halved between 1920 and 1950. The percentage fell further in the '50s, when most of the foreign-born were in later stages of their lives.[43] The slowing of immigration starved ethnic culture. The number of ethnic theaters and musical organizations decreased after the Second World War. Social halls entertained only the older people. Ethnic newspapers folded; although a few survivors still provided information on the old country, the local churches, and the social affairs of the particular group in question, the proliferating English-language weeklies reported on their retail districts, displayed ads for downtown stores, and supplied information about educational opportunities and social-service agencies.[44]

While the lack of fresh immigrants weakened the cohesion of what a writer called the "nationality groups," the neighborhoods gained a new kind of coherence during the '40s and '50s. Fewer arrivals and departures allowed people and cultural patterns to settle into place. Neighborhoods also became more tightly bound to the rest of the city. Every home, block, and retail street, in other words, was becoming like the others, as English became more widespread, ties to the old country frayed, and more people adhered to the rules, norms, and culture of American society. The gathering uniformity was a regular topic of books and magazine pieces. It was quickened by a federal government that expanded its role in the lives of city people, by the experiences of millions of second-generation boys in World War II, and by a postwar economy that featured a huge manufacturing sector, a big union movement, and an expanding mass market.[45]

Alfred Kazin was part of this finer blending of what Barzun called the "grand mixture." Kazin grew up in a Jewish neighborhood in the 1920s and 1930s. "We were of the city," he wrote, "but not somehow in it. Whenever I went off on my favorite walk to Highland Park in the 'American' district to the north, on the border of Queens, and climbed the hill to the old reservoir from which I could look straight across to the skyscrapers of Manhattan, I saw New York as a foreign city." As a boy, he could not know that Manhattan and his Brooklyn neighborhood were, as he later put it, "joined" in him. But by the '40s he was not only "of the city" but also fully "in it," and so could write *On Native Grounds,* a now-classic account of the history of American literature. Kazin wrote not about Jews or the immigrant experience, but about the greatest of American writers. In so doing he went from being an ethnic outsider living in a Brooklyn neighborhood to being an insider living in Manhattan and writing about American literature. In 1951, Kazin wrote about his childhood community as the distant past. He did so, in part, because many of the Jews had left and taken with them much of the Jewishness of the place. He did so, as well, because the old neighborhood's people and culture were being folded into the blended American life that he personified.[46]

Kazin's story resembles that of millions of second-generation immigrants who came of age in cities during the '30s, '40s, and '50s. They had ties to Europe through their parents, grandparents, and aunts and uncles, but they were homegrown. They grew up in American cities that for the first time were not significantly influenced by recent newcomers from Europe. They had fought a monumental war across two oceans. So even if, as a writer put it in 1955, the "nationality groups remain well organized around a persistent, intense sense of identification that spans the gulf between generations," the ethnic "ghettos" of American cities had clearly "shrunk" since the war.[47]

They shrank in population: many older immigrants retired or died in the '50s, and second- and third-generation Americans took their places. They shrank culturally, too. The folk songs of the old country, for example, were losing out to Bing Crosby, Frank Sinatra, Ella Fitzgerald, and the big bands. What was left of ethnic theater had little chance against *Dragnet, The Honeymooners,* and *I Love Lucy.* Foreign-language newspapers lost their audiences to *Life* and *Reader's Digest.* Everyone who grew up in America during the '30s, '40s, and '50s grew up on baseball and the movies. Even though delicatessens sold ethnic dishes and some mothers cooked meals from the old country, more and more people ate chops and meatloaf, drank orange juice that had been frozen, and prepared their food from cans and boxes. So while some stores and eateries retained "the flavor of their homeland," it was observed that "many others strive to become modern and meet the demands of a new generation." Even the churches lost some of their cultural identity. Most were built between the late 1800s and early 1900s and were still fixtures in the neighborhoods. Some had even grown larger and richer after the war. But the inveterate sectarianism of most Protestant denominations blended into a larger brand of Protestantism. The patron saint and national origin of a Catholic parish meant little to native-born Americans. Fewer services were conducted in Greek, Polish, German, Latin, or Hebrew.[48]

American culture has been built on the assimilation of newcomers. But assimilation quickened and deepened in the 1940s and 1950s, when city neighborhoods became more settled and cohesive than ever. When in the city, remembers Newark native Philip Roth, you might have thought of yourself as a Newark Jew, a Newark Pole, or a Newark Italian, because those designations still meant something in the neighborhoods. But once outside the neighborhood, and certainly once outside the city, you were just an American. Saul Bellow caught the mood, in 1949, with his opening line in *The Adventures of Augie March:* "I am an American, Chicago born."[49]

The neighborhoods were settled down even further by a manufacturing economy that pulled its large, well-paid, and highly organized working class right into the center of American life. Unions wielded tremendous clout in the 1950s. Blue-collar workers spent their rising paychecks on the same products and amusements that white-collar employees and even professionals enjoyed. The American worker, crowed union boss George Meany, "has never had it so good." He was "the best-paid, best-clothed, and best-housed worker in the world."[50]

This prosperous and stable working class was rooted in the neighborhoods. Although manufacturers built new plants in suburbs, and some workers moved there, most blue-collar jobs remained in cities during the '50s, and over half of all employed people living in cities were blue-collar workers. They worked in small shops scattered throughout residential districts, in big plants surrounded by blocks of working-class housing, in old warehouse and manufacturing districts along waterfronts and rail lines, and in big factories erected during the 1920s on what was then the edge of the city. They made cars, motors, machine tools, and grinding wheels; manufactured chairs, tables, lamps, and light bulbs; mass-produced spaghetti, bread, candy, and cookies; canned meat, fruits, nuts, and vegetables; marked, cut, and stitched cloth; and moved those products on trucks and trains and through docks and warehouses.[51]

Urban manufacturing was enjoying the last leg of a long run after the war. It is true that New England's textile and shoe mills hit their peak in the '20s, that some of Detroit's older automobile plants closed in the '50s, and that Philadelphia in the '50s had pockets of "industrial blight," full of what a newspaper described as "crowded, crumbling, dingy and scabrous structures, many of which could not stand rigorous enforcement of fire and health regulations."[52] By and large, however, the industrial districts of the nation's cities were intact and served as the economic base of many neighborhoods. Older workers figured on retiring from the plant they had worked at for years. Younger men anticipated long careers as industrial workers. Store owners depended on factory wages. No one imagined, in 1955, that in just five or ten years urban manufacturers would trim their work forces, go out of business, move to suburbs, or build new plants in Mexico. San Francisco was typical. Its manufacturing sector, and its blue-collar neighborhoods, held steady through the '50s. During the next decade, the city lost two hundred manufacturers, fifty-eight hundred factory jobs, and the wages that had supported thousands of families and hundreds of neighborhood merchants.[53]

While the prosperity of blue-collar workers rested upon the industrial might of the nation, it took unions to pry wealth from manufacturers and give it to workers in the form of higher wages, pension funds, and stable employment. Unions represented nearly three-quarters of industrial workers by the middle of the '50s. That included seventeen million workers—about 30 percent of the entire labor force and considerably more of the urban work force. Unions organized nearly every large plant and many of the smaller ones. At no other time did city neighborhoods have so many union men, and at no other time did those working men do so well: the average income of an industrial worker rose almost 50 percent between 1945 and 1960. Never had the commercial streets of blue-collar districts, and of mixed blue- and white-collar neighborhoods, appeared more crowded or prosperous. Union men now thought of themselves more as middle class than as working class, more as Americans than as Hungarians, Italians, Poles, or Germans.[54]

Unions gained respect after the war in spite of crooked teamsters, a number of corrupt locals, and plenty of factional fighting. Around 75 percent of the population, and over 70 percent of businessmen, said they "approved of labor unions" by the middle of the '50s. But unsavory material turned up at hearings on union corruption, and teamster president James Hoffa admitted he had a police record "maybe as long as your arm." When the AFL-CIO expelled the teamsters at the end of 1957, almost half of the public believed graft was rife in labor unions. Close to two-thirds, however, still approved of the unions.[55]

Unions joined the establishment in the '50s. Labor leaders were welcomed on civic boards, community chests, and patriotic organizations. Scores of union officials served as labor attaches in American embassies. Teamster boss Dave Beck addressed chambers of commerce and Rotary clubs all across the nation in the early '50s; he was also a member of Seattle's Civil Service Commission, the Washington State Prison Board, and the Board of Regents at the University of Washington. Union halls in city after city, reported the *Saturday Evening Post*, moved from "dismal labor temples on side streets to well-heeled midtown respectability."[56] Walk around old factory districts or working-class neighborhoods in our time, and you'll see small union halls built in the '50s for local electricians, welders, or pipe fitters. You'll see imposing glass-and-steel halls built by the mighty unions of the automobile and steel workers. You might see a version of San Francisco's grandiose Sailors' Union of the Pacific building, which was erected on Rincon Hill in 1953 to overlook a still thriving waterfront.

Many workers took pride in their work and their union. Manhattan's garment shops, for example, had a vigorous union culture of nearly guild-like relationships between petty capitalists and their cutters and sewers. Workers in metal, machine-tool, and engine-making factories thanked their unions for safer working conditions. All sorts of workers used their local union halls as gathering places and watering holes and enjoyed playing on union softball and bowling teams. Nearly every worker understood, at some level, that he owed his union for the prosperity of his family and his neighborhood.[57]

Despite all that, American unionism did not translate into a strong, widespread, working-class identity—not even in neighborhoods full of blue-collar workers. "Radicalism has vanished from the union move-ment," says an older union man. "All the guys talk about is sex, baseball, and cars." And why not? Union men in the '50s made more, on average, than white-collar workers. The big union battles had been fought in the '30s, and most of the great strikes were waged right after the war. The av-erage worker readily endured the few strikes of the '50s because he had some savings, got credit from local merchants, and received money from the strike fund.[58] The fund was typically managed by unions that had be-come huge bureaucracies with little interest in hearing from the rank and file. Warner Bloomberg was a writer and steelworker in Gary, Indiana. He lamented, in 1955, the "rank and file workers' loss of an image of lead-ership with which they could identify and in whom they could place great trust." Small wonder, wrote a journalist who covered the labor move-ment, that most workers seemed to "think of the union as an instrument for delivering them a regular raise in pay and for talking up in their be-half if they get in trouble with the boss. That is the beginning and end of it." George Meany, boss of the AFL-CIO and its fifteen million members, stated his position in 1955: "We believe in the American profit system. . . . Our goals as trade-unionists are modest, for we do not seek to recast American society in any particular doctrinaire or ideological image. We seek an ever rising standard of living. Sam Gompers once put the matter succinctly. When asked what the labor movement wanted, he answered 'More.'"[59]

If the typical working man did not have a deep philosophical attach-ment to his union, neither did he have a deep identity as a worker. That identity was diluted by the house and the car, by the young men who sim-ply signed a form to join the union, by collective bargaining agreements hammered out in downtown office suites, and by the demise of radical politics in the late '40s. "Less than ever before," wrote an observer of the working classes in 1953, "does the laborer identify himself with his job, think of himself, as an individual, primarily in terms of his position at a particular type of work in a particular plant. The riveter, the laster, the crane operator sheds the marks of his employment when he leaves the shop. The nature of his work will only slightly influence his life outside."[60] An onlooker at a Labor Day parade on New York's Fifth Avenue guessed the average age of marching workers to be around fifty; the younger guys, she supposed, preferred to stay at home. If most union men did not march in public, neither did their union membership carry over into pol-itics. Forty percent of manual workers and many union men voted for Eisenhower in 1952. More than 70 percent of manual workers applauded the job Eisenhower did in 1953. Almost 60 percent of union members, and nearly 70 percent of their wives, approved of the president in 1954.[61]

The great success of the unions, the prosperity of the neighborhoods, and the dilution of ethnic culture meant that the native-born worker, according to the author of *Workingman's Wife,* "feels much more at home in his society, and he feels more securely that it *is* his." Class and ethnicity, which for so long had been the markers of neighborhood identity, gave way in the '50s to a stronger blend of "Americanism," which meant an identification with a very coherent national culture and a sense of belonging to an increasingly roomy middle class. Many city neighborhoods, in other words, housed union men who were stripped of ethnic culture, except for heritage, and freed from an identity rooted firmly in class.[62]

In many ways, those neighborhoods were fine arrangements for living. The stoop, the block, and the retail street, along with the church, school, and union hall, made an intimate and nearly complete little world that had a certain stability about it. But if residents were often kind and generous, they could just as easily spread gossip, resent a neighbor's success, believe women belonged in the house, mistrust those who dressed differently, suspect an honest Jewish merchant of cheating them, despise the upstairs couple for fighting on Saturday nights, and hector the black family across the street. All that familiarity, in other words, bred as much contempt as it generated respect. The sense of belonging, which so many people remember so fondly, also meant knowing your place—or paying a price.

Despite its faults, the typical neighborhood did provide its residents with security, familiarity, and stability. A second-generation Irish Catholic, whose father was a mailman, wrote the following about a part of San Francisco's Mission district:

We were dominated completely by family and church and we were absolutely secure. Every one of our relatives from both sets of grandparents to each of our many cousins lived within walking distance of each other's houses. . . . Our neighborhood was our world. Although there were occasions when we took the J Streetcar or the 26 or 14 bus downtown, we rarely visited other districts of The City. . . . Our church and school were only a few blocks away and nearby Mission Street offered complete shopping and entertainment. Charley, the grocer on the corner of 26th and San Jose Avenue who gave us end pieces of bologna and salami before he shooed us out the door, had been on the same corner when my father was a boy. There was an overpowering sense of continuity. We were "Wishie's kids" or "Florence's girls" without need for further identification.

Such images of neighborhood life explain why 125 people who grew up in a residential district in Buffalo attended a *neighborhood* reunion in 1984—some twenty-five years after they had moved out.[63]

The Black Metropolis

American cities have always been infused with the pluck, hopes, and ways of outsiders. The pattern has been for the latest arrivals to move into low-grade houses and apartment buildings, vacated by earlier newcomers who had moved up the social hierarchy by moving out of the old neighborhood. Almost all of those outsiders were Europeans, until the 1920s, when southern blacks started moving in large numbers to New York, Philadelphia, and Chicago, and in smaller numbers to cities like St. Louis, Cleveland, Newark, Baltimore, Kansas City, and Detroit. Migration from the South nearly ceased during the Depression, but it resumed in the 1940s and 1950s, this time in much greater numbers and to almost every big city.

Blacks were not the only newcomers to cities during and after the war. A sizable group of white, Protestant, old-line Americans moved to Nashville, Kansas City, Memphis, St. Louis, Indianapolis, Chicago, Baltimore, Louisville, Detroit, Pittsburgh, and Cincinnati. To move from Appalachia, the Ozarks, and the backwoods of Kentucky meant adjusting to traffic, to the quick pace of work, to being called a "hillbilly." Migrants tended to live in slums; they wore their poverty on their backs and blared country music from jukeboxes in local bars and hash houses. A magazine piece called "The Hillbillies Invade Chicago" shows how most established residents probably viewed these newcomers:

Settling in deteriorating neighborhoods where they can stick with their own kind, they live as much as they can the way they lived back home. Often removing window screens, they sit half-dressed where it is cooler, and dispose of garbage the quickest way. Their own dress is casual and their children's worse. Their housekeeping is easy to the point of disorder, and they congregate in the evening on front porches and steps, where they find time for the sort of motionless relaxation that infuriates bustling city people.[64]

More than half a million Puerto Ricans moved to the mainland during the '40s and '50s. Some went to Chicago and Philadelphia. Others rented apartments on the edge of Newark's central business district.

Most settled in dense Manhattan apartment buildings in the southeast corner of Harlem, once home to Italians and Jews.[65] Or they found rooms and apartments in a transitional area west of Central Park, where brownstones became rooming houses into which Puerto Ricans, according to a sympathetic if resentful observer, "crowded ten in a room, spilling over the stoops into the streets, forced through no fault of their own to camp indoors as well as out, warming their souls on the television set and blanking out thought with the full volume of radio." The same happened higher up on Manhattan's West Side. Landlords turned apartment buildings into hotels that rented rooms by the week or month to Puerto Ricans, blacks, and poor whites.[66]

While Puerto Ricans and Appalachians altered the character of some blocks in a few cities after the war, the arrival of blacks brought big changes to many cities. The movement of blacks to northern cities happened quickly, in large numbers, and against the wishes of most whites. In 1940, blacks made up less than 10 percent of the population in most large cities, and most black communities numbered only in the thousands or few tens of thousands. Major exceptions were Philadelphia, which had close to two hundred thousand, and Manhattan and Chicago, which each had around a quarter-million. Whatever the starting point, the black population in most big cities nearly doubled during the '40s and '50s.[67]

The growth in population meant an expansion of territory. In Chicago, which was an extreme case, nearly four thousand apartments, flats, and houses went from white to black each year between 1945 and 1949, a rate that nearly doubled to seventy-five hundred during each year of the '50s.[68] It was usually but a few years between the arrival of the first black on the block and the exodus of almost all of the whites. A white couple told me about living, and wanting to live, in racially mixed neighborhoods in Detroit. But they had to move twice in ten years because of the rapid transition from mostly white to nearly all black. Detroit's Dexter Avenue area, reported *Commentary* magazine in 1956,

is an excellent example of one in transition. On Dexter itself are all the identification marks of a Jewish neighborhood—the kosher butchers, the delicatessens, the bookshop selling literature in Hebrew and Yiddish, the Barton's candy shop; on neighboring Linwood is a bagel bakery and a caterer who deals in cocktail knishes. The residential streets that run off these two main avenues, lined for the most part with two-family homes, are abloom with "For Sale" signs; most of them are already mixed streets. Just as the Eastern Market, which always mixed Michigan garden truck with its Italian products, now has stands for collard greens, so the signs on the grocery windows begin to indicate that the neighborhood is changing from Jewish to Negro. Farther east, on Twelfth Street, which was once one of the Jewish main streets, there are still a few relics of the past, like one popular delicatessen-type restaurant, which despite the fact that it is almost a Jewish island in a Negro sea, is reluctant about serving Negroes at its tables.[69]

So it was in every big city. Before 1950, Euclid Avenue marked the border between blacks and whites on Cleveland's east side. During the '50s, blacks crossed that boundary and entered the Hough neighborhood. Whereas blacks had occupied only 7 of 1,350 dwellings in a 21-block area of Hough in 1950, 10 years later the area housed over 1,000 black families. In 1950, the Matsen-Southhampton neighborhood of Buffalo was almost all white; by the end of the decade it was pretty well black. Bedford-Stuyvesant in Brooklyn was nearly all white during the war, was half black in 1950, and was mostly black by the early 1960s. Blacks moved in and whites moved out of the Crown Heights, Brownsville, and Bushwick neighborhoods of Brooklyn as well. In the Corona section of Queens, the black population doubled during the 1940s, accounting for 25 percent of the population by 1950. It doubled again during the 1950s, to comprise half of the population by 1960. The North Lawndale section of Chicago had eighty-seven thousand residents in 1950. Nearly all were Jews. By 1960 there were eleven thousand Jews and a hundred thousand blacks. The same thing happened on Chicago's South Side, in parts of Newark's central and west wards, and, on a smaller scale, in San Francisco's Fillmore district, West Oakland, the Homewood-Brushton area of Pittsburgh, and parts of North Miami.[70]

Here is how it happened in North Philadelphia. In 1949, whites occupied each of the ninety row houses on the block of East Hortter Street that runs off Germantown Avenue. Four years later, whites lived in just thirty-two of those houses. The change started in late November 1949, when a black family moved into a house on the block. "It was a panic," one of the remaining white residents recalled in early 1954. "Groups of owners got together in living rooms and swore they wouldn't sell to Negroes. Older residents, especially, feared property values would drop to nothing. There was uproar and bedlam and a lot of foolish ill-feeling. Some hot-heads talked about ganging up to keep Negroes out. We talked them out of it." The residents made a pact not to sell to blacks. It lasted a month. "It was every man for himself then," a resident recalled. "The block was plastered with 'For Sale' signs on New Year's Day, 1950. It looked like the Chicago wheat pit. These real estate fellows were racing each other to homes to get listings. Some of them even had a fist fight and we had to call the police. They had their pockets full of cash, agreements and receipts for anyone they could pressure into selling fast."[71]

By New Year's Day, 1951, blacks lived in more than half of the houses on the block. Many of the rest were for sale. The buyers, according to a newspaper report, "were Negro veterans, Negro government workers, school teachers, office workers and some professional men and women." Whites living on the block in 1954 apparently saw their new neighbors as good neighbors: the kids played together and blacks maintained their houses as carefully as whites did. The motivations for leaving, or for thinking about leaving, must have ranged from virulent prejudice to concerns about property values to not wanting to be the only white family left on the block. One of the remaining white residents said this in 1954:

[Negroes] are looking for better places to live, and they have a right to them. None of us blame them for that. But I prefer to live among whites—and I presume most Negroes would prefer to live among Negroes. Call me prejudiced if you like, but a lot of people must feel the same way or why would so many panic when a Negro family moves into their block? It seems to me we have to face things squarely and decide which it's to be. Shall we improve the areas where the Negroes already live so that they'll be happy to stay there—or shall we, by indifference to their problems, encourage them to move in anywhere and everywhere they can find a house?[72]

That was a fairly compassionate view for the time. The general attitude of most working- and middle-class whites may have been described by a liberal member of the Philadelphia Catholic Housing Council, who, after talking with a neighborhood group about the changing racial composition of the city, recorded this in his diary: "Silence. Stone cold silence. The people were just intensely and silently aggravated by the movement of Negroes throughout the city. As I left I knew that they despised me because they believe I am 'against them.'"[73]

The transition was not violent on Hortter Street, but it was in other sections of North Philadelphia. It was violent in many parts of Chicago, too, where most of the problems between blacks and whites occurred along shifting racial borders. Attacks on individual blacks, assaults on black-owned houses, and group protests against blacks moving into white neighborhoods took place primarily in working-class districts inhabited by Catholics. Whites in those areas may have been less tolerant about racial integration, fearing a loss of status if they were to live among blacks. Many of those whites could not afford to sell and buy elsewhere, and so were especially vulnerable to a fear of declining property values.[74] Everyone in Chicago, and many Americans living elsewhere, heard about events in Cicero, a working-class suburb built on the western edge of the city in the 1920s. In July 1951, a black city bus driver and graduate of Fisk University tried to occupy an apartment he had rented. Governor Stevenson sent in five hundred national guardsmen to quell two nights of rioting by a mob of whites. The police arrested more than a hundred persons. *Life* reported the incident as a "disgrace for Cicero."[75]

Even a peaceful transition from a white to a black neighborhood still hurt. The Corona district in Queens had a small black population before the war. More blacks moved in during the '40s and '50s. "White people starting moving away," a woman recalled, "they moved away from Corona. Because the associations were bad. It was *black*, it was ghetto, people didn't have jobs."[76] Only one of those associations was true: it was black; but it wasn't a ghetto and the people did have jobs.

KEEPING IT TOGETHER

In their 1961 postscript to *Black Metropolis*, the most comprehensive book ever written about black life in urban America, St. Clair Drake and Horace Cayton depicted Chicago's South Side as essentially the same place they had written about in 1945: "Bronzeville, as a compact and self-conscious community, still exists. There has been very little change in the fundamental values of its people or in the basic social structure and cultural patterns of the community."[77] Nearly every other black neighborhood was also a compact and self-conscious community in the '40s and '50s.

At the base of this community was a great variety of incomes and occupations, interests and tastes, southern and northern experiences. A 1956 magazine article described Detroit's Paradise Valley as a "Negro slum." It did have "collapsing frame houses, with sagging porches and window frames askew." It also had attractive two-family houses and nice apartment buildings. Such extremes represented what a member of the Urban League saw as the essence of the community: "a mixture of everything imaginable—including overcrowding, delinquency, and disease. It has glamour, action, religion, pathos. It has brains and organization and business. Not only does it house social uplift organizations, but it supports the militant protest groups."[78] A mixture of everything imaginable: that was the trademark of the black neighborhood in the '40s and '50s. Established professionals and businessmen shared the retail avenue with recent migrants from the South. A pawn shop and a seedy bar stood between a spick-and-span Woolworth's and a fine jewelry store. A Negro press tried to appeal to all blacks. Such a neighborhood, in other words, was a black metropolis unto itself.

Despite a pocket of wealth here and a cluster of poverty there, the confinement of blacks to certain parts of the city prevented the housing market from efficiently sorting residents by income and status. Consider the situation of the black upper classes. It was true, as *Ebony* magazine reported, that "Harlem's most talked-about men and women in law, sports, civil liberties, music, medicine, painting, business and literature live on Sugar Hill." A former resident of Sugar Hill remembered coming home at night from midtown, where she worked as an actress in the early '50s: "At one in the morning, I would be taking the Eighth Avenue subway up, getting off at the top of the Hill. I had no fears whatsoever. It was a benign community—a gentle, warm community, where people cared about each other, with great good will and politeness." It was also true that hundreds of working-class people lived on elite Sugar Hill. Many of them took in boarders, and some even earned rent money by charging for apartment parties that featured live music.[79]

Chicago's black professionals and entrepreneurs tended to cluster in the southeastern part of the so-called black belt. But what the authors of *Black Metropolis* wrote in 1945 was still true ten, and even fifteen, years later: "Instead of middle-class *areas* Bronzeville tends to have middle-class *buildings* in all areas, or a few middle-class blocks here and there." Brooklyn's Bedford-Stuyvesant district had many tree-lined streets with handsome row houses and apartment buildings; it also had many classes of black people living in close quarters. Well-off blacks did tend to live along the "Black Gold Coast" of Queens's Ditmars Boulevard, and black entrepreneurs and professionals resided in the Boston-Edison neighborhood of Detroit, in the Hunter and Collier Hills sections of Atlanta, and in parts of West Newark. On the whole, however, wealthy blacks could not congregate themselves as thoroughly as their white counterparts.[80]

Just as upper-class blacks tended to reside in certain sections and blocks, so did the desperate and destitute cluster in specific areas. Poor people congregated in places like the northern part of Chicago's South Side, the eastern fringe of San Francisco's Fillmore district, and hard by the central business districts of Newark and Washington. Many had just arrived from the South. They rented rooms in dilapidated brick row houses and three-story Victorians that had been converted to tenements, or they lived in apartment buildings that had been cut up into single-occupancy rooms. The commercial corners featured tiny liquor and grocery stores, pool halls and beauty parlors, pawnshops and store-front churches. The great numbers of southerners coming into such areas, and the high turnover of rented rooms and efficiency apartments, altered the composition of classrooms so rapidly that a teacher would have in June but a few of the pupils who had started in September.[81] Some run-down sections concentrated real distress, like a section of Brooklyn where a journalist counted "sixteen store-front Pentecostal churches in some blocks, separated here and there by a bar or liquor store. Here you can see the prowl cars slowly moving down the streets, the refuse spilling into the gutters, the gleam of flattened beer tins on the pavement, the rotten smell of decaying buildings, the vacant store windows blindly facing dingy streets."[82]

Despite some geographical clustering by income, however, the various social classes lived among one another and came together on lively commercial and entertainment strips that served as a kind of downtown for the black metropolis. Central Avenue in Los Angeles, Fillmore Street in San Francisco, Pennsylvania Avenue in Baltimore, Seventh Street in West Oakland, Center Avenue in Pittsburgh, 125th Street and Lenox Avenue in Harlem, parts of Forty-seventh Street, Sixty-third Street, and South Parkway in Chicago—such streets gave identity to their communities. Recent arrivals and ordinary working people shared these main drags with local big shots. A man still living in West Oakland told me about shining the shoes of dock workers as well as lawyers on a busy Seventh Street in the early 1950s. A person of modest means could boast, or at least tell a plausible story, about meeting a celebrity on the central avenue: "Sometimes I run into Duke Ellington on 125th Street and I say, 'What you know there, Duke?' Duke says, 'Solid, ole man.' He does not know me from Adam, but he speaks. One day I saw Lena Horne coming out of the Hotel Theresa and I said, 'Hubba! Hubba!' Lena smiled."[83]

Five or six blocks of San Francisco's Fillmore Street served as the neighborhood's hub of commerce and entertainment. Among dozens of establishments were a Woolworth's, an Owl Drug Store, a Thom McCann shoe store, Diller's clothing store for women, Deb's Department Store, a Bank of America, Louie's Haberdashery, and many fine nightclubs.[84] The street remained in good shape throughout the '50s. When the city council granted permits in 1952 for a pawnshop and a second-hand store, the Fillmore merchants' association complained in the neighborhood paper that "it would greatly deteriorate the quality of the Fillmore as a shopping district" if the stores were allowed to move in. That reaction showed their pride in the street. It may also have betrayed their fear of the street's eventual decline. For even in its heyday, from the late '40s to the late '50s, when Fillmore Street was said to be "bustling with vigor," it served the diverse clientele of a relatively poor neighborhood. Like black main streets everywhere, Fillmore was a mix of fancy stores at the high end, raunchy dives at the low, and almost everything else in between, including a five-and-dime, donut shops, and back rooms that served as betting parlors.[85]

Fillmore Street had a thriving jazz scene in the '40s and '50s. The weekly *Sun Reporter* announced and reviewed performances at the neighborhood clubs, which attracted every local player and plenty of big names, too, from Miles Davis and Thelonious Monk to Count Basie and Duke Ellington. "You might have four clubs in a block, two on each side of the street. And then you go around a couple more blocks and then you have another couple of clubs," explained drummer Earl Watkins.

You had the Club Alabam, which was one of our old established jazz clubs. Across the street was the New Orleans Swing club. . . . On Fillmore between Sutter and Post, you had Elsie's Breakfast Club. . . . Then down the block was the club called the Favor. Across the street from that was the Havana Club. And then when you went down the next block, Fillmore between Post and Geary, you had the Long Bar, which had Ella Fitzgerald. Then down another couple of blocks and you had the Blue Mirror. Then across from the Blue Mirror, they had the Ebony Plaza Hotel. In the basement, they had a club. And if you went up Fillmore to Ellis Street, you had the Booker T. Washington Hotel. And on their ground floor, in the lounge, they had entertainment.

"The Fillmore used to swing," remembered another musician. "During the heyday we could go to work Friday night and not get off work until Monday morning. We never stopped. Jam sets would go on for hours. We went from club to club, playing." It was a real social scene, too, for whites as well as for blacks. "Everyone got dressed up in those days," remembered singer Sugar Pie DeSanto, a cousin of Etta James. "We really got sharp."[86]

So it was in every black neighborhood. Jimmy Witherspoon, for example, remembered Central Avenue in Los Angeles after the war: "We had so many nice night spots," from full-scale cabarets and big ballrooms to "blues incubators like the Barrelhouse in Watts and after-hours spots all over the place. . . . No wonder such great music came out of Central Avenue." There were clubs all over the South Side of Chicago, like the Regal and Savoy on Forty-seventh street, the musicians' jam joints on Garfield, and the Sutherland Hotel on Forty-seventh and Drexel.[87] Dizzy

Gillespie, Miles Davis, John Coltrane, Nancy Wilson, Earl Hines, and Duke Ellington all played and stayed at the Sutherland. "When I was a kid growing up along 47th Street," recalled a jazz player, "everyone knew about the Sutherland, everyone played there, all the great musicians hung out there." A former jazz disk jockey remembered it as "a hangout—it was *the* hangout. And on weekends, it was about a buck, a buck and a half to hear a show. Imagine going to hear John Coltrane for a buck on a Saturday night. Unbelievable." The place seated about a hundred and broadcast live shows throughout the late '50s.[88]

Black musicians everywhere grew up in neighborhoods that pulsed with jazz in the '40s and '50s. They had ready audiences for their music and every block had at least a few knowledgeable jazz fans. Young musicians listened to neighborhood talents in side-street bars, and they saw, and occasionally sat in with, the great performers in clubs on the big avenues. "Then a musician was really getting his music across," recalled Chicago's Andrew Hill, "because there were bars on every corner and you couldn't go anywhere without hearing music. That's why people were so into music." If you weren't listening to jazz bands in local clubs, you still heard the music because "all the jukeboxes, they had jazz," remembered saxophone player Clifford Jordan; "but nobody called it 'jazz' then. It was just music. It was just our music, folk music."[89]

Jazz is an urban music, writes a critic, and "like everything else in the city, refuses to stand still. It's always looking for ways to outdo itself, and is sustained in its search by the rich ground of black urban life."[90] That describes perfectly the development of jazz from the 1920s, when it assumed its modern form, right on through the 1950s, when it enjoyed its last great creative burst. Jazz evolved in those decades as musicians experimented with their material, competed with one another, and felt a close and creative tension with their audiences. Black urban life was especially rich in those days because the neighborhoods combined the different social classes along with southern and northern experiences. Jazz expressed the mixture, and so it portrayed, as well as anything could, the experience of the entire black metropolis.

The great mix of rich and poor, of newcomers and old-timers, of educated and illiterate, of jazz players and jazz fans—all rubbing shoulders on the major commercial streets—was made coherent by a dominant fact of life: these were *black* communities. The neighborhood weeklies wrote almost exclusively about black life. They covered the national sport stars, political figures, and the work of the NAACP. They also reported on local church activities, crime in the community, and the building of public housing.[91] It made sense for the weeklies to cover black life. The daily papers paid it little attention. And even though income, level of education, and family histories divided blacks just as they did whites, such differences could not express themselves geographically in black communities and were often overridden by the unifying experience of race. That is not to say most of the people spent most of their time thinking about racial identity or pursuing social justice. But being black meant always having to deal with race in one way or another, whether it was a fear of the police, being watched in downtown stores, facing overt hostility from certain whites, being unable to eat in some restaurants, or just living in a black neighborhood. So the focus on black life by the weekly papers struck a chord with almost everyone in the community.

While the weekly paper meant to represent the whole community, it also reflected its divisions. Each weekly had society pages, for example, which announced balls and described the weddings and anniversary celebrations of its upper classes. Nicely dressed women appeared in photographs taken at their clubs and social events. Newspaper columns like "Flo Sezs" and "Connie's Corner" spread polite gossip. The *Pittsburgh Courier* reported on the professional and political activities of educated women in a column called "A Woman's World." Although most editorials addressed topics of national importance, the opinion pages occasionally reflected local concerns of the city's black upper classes. They worried, for instance, about street crime, back-room betting parlors, even loud behavior in lower-class churches. They also declared upper-class pride: "It is our impression," wrote the Fillmore's *Sun Reporter,* "that the percentage of Negroes who attend opera, symphony, and similar events, would compare, favorably, man for man, with the ratio of white population which subscribes to these events." Every weekly also dedicated a couple of pages, like the "Pews and Pulpits" section of the *N.Y. Age Defender,* to the activities, services, and outstanding members of local churches. The churches were vital parts of the neighborhood. Numerous store-front churches on side streets offered Sunday services for the lower classes. Most cities had a version of Chicago's Greater Harvest Baptist, which, according to the *Chicago Defender,* provided "the greatest service to the needy of any non-professional organization in the city." Greater Harvest served more than ten thousand free meals under its roof in 1956, and gave clothing to thousands more.[92] But the wealthier, well-organized, and more sophisticated churches got almost all of the newspaper coverage.

The educated and the wealthy comprised a fairly compact lot, but they had their own divisions, too. There were various clubs and churches, disagreements over political strategy, conflicts between older families and newer ones, even hierarchies by shades of color. The Fillmore's *Sun Reporter* aired a typical spat. The "Twentieth Century Girls" wanted to rent a hall from the Sailors' Union of the Pacific for a social event. A newspaper columnist called the union racist and complained that several of our "clubs have betrayed the cause of their own working men by putting their dollars into the coffers of men who brazenly admit their contempt for Negroes." It would be better, he wrote, if the social clubs "became adult organizations and took part in community betterment instead of living in make-believe dreams of self-centeredness and importance."[93]

Or consider Berry Gordy, who came from a well-off family in Detroit but who lost any chance of getting into the city's inner circle of wealthy blacks when he opened Motown in 1959. For he now consorted with the coarser classes.[94] Or a woman who moved to the better part of the black neighborhood in Queens: "If you didn't have a college degree," she remembered, the neighbors "would walk all over you. I was surrounded by people who owned their own businesses, and they got to putting on airs. I told them, 'I pay the same taxes as you do, and I go to the bank just like you, I throw out just as much garbage as you do, and I do just as much for my children as you do.' That stopped them!"[95] That the upper classes fought among themselves, and distanced themselves from the lower classes, was normal and nothing to decry. But the black metropolis badly needed firm leadership.

Providing such leadership was difficult, in large part because most well-off blacks were professionals rather than businessmen. The "black bourgeoisie," wrote E. Franklin Frazier in 1957, does not therefore "exercise any significant power within the Negro community as an employer of labor. Its power within the Negro community stems from the fact that middle-class Negroes hold strategic positions in segregated institutions and create and propagate the ideologies current in the Negro community." Black entrepreneurs, according to the authors of *Black Metropolis,* "have not broken into the industrial and commercial 'big time,' and the bulk of the Negro market's money goes to white banks and loan companies, the supermarkets, the downtown department stores, and to white insurance companies—not to Negro enterprises." Most black businessmen ran small trades in areas where white people seldom bought.[96]

The economic weakness of the upper classes limited their ability to employ local residents or provide them with goods and services. The middle classes, accordingly, had fewer chances to build entrepreneurial skills or sell professional services to businesses. Poor people had a hard time find-ing decent jobs in the neighborhood. The lack of economic links between the classes translated into weak social connections. Local businesses, for example, rarely sponsored Little League baseball, participated in com-munity development projects, or lobbied city hall for better sidewalks and bus service on the major retail strips.

The economic weakness of the middle and upper classes also meant that outsiders owned many businesses in the community. Locals com-plained that outsiders drained money from the neighborhood, that white shopkeepers overcharged black customers, and that too few blacks worked in the stores. Most black businessmen felt helpless to correct the situation. "One of these days," wrote San Francisco's black newspaper, "some reputable merchant from outside this area will come in and set up a branch staffed by Negroes."[97]

The control of many local businesses by outsiders incited resentment, especially against Jews. James Baldwin may have exaggerated, but he re-vealed a truth when he remembered

meeting no Negro in the years of my growing up, in my family or out of it, who would really ever trust a Jew, and few who did not, indeed, exhibit for them the blackest contempt. On the other hand, this did not prevent their working for Jews, being utterly civil and pleasant to them, and, in most cases, contriving to delude their employers into believing that, far from harboring any dislike for Jews, they would rather work for a Jew than for anyone else.[98]

The two groups met regularly in the neighborhoods, in part because blacks often moved into Jewish areas, in part because Jewish merchants and landlords saw business opportunities in black areas. Either way, Jews owned property, ran stores, competed with black businessmen, and dealt with black customers.[99] A version of this scene, from Edward Wal-lant's *The Pawnbroker*, must have played out millions of times in black neighborhoods:

"I can only give you two dollars," Sol said, flipping over the pages of his ledger, looking for nothing in particular. "You've left an awful lot of things lately, haven't redeemed anything."

"Aw I know, but Mistuh Nazerman! Why, my goodness, these candlesticks is very high quality, costed twenny-five dollars new." She chuckled indignantly, shook her head at his offer. "Why I could get fifteen dollars *easy* down at Triboro Pawn."

"Take them to Triboro, Mrs. Harmon," he said quietly.

Mrs. Harmon sighed, still shaking her great smiling face as though in remi-niscence of an atrocious but funny joke. She clucked through her teeth, shifted heavily from one foot to the other. Her dignity, that much-abused yet resilient thing, suffered behind her rueful smile. . . .

"Les jus' say five dollars the pair and forget it, Mistuh Nazerman," she said, breathing hopefully on him.

"Two dollars," he repeated tonelessly, frowning over a name in the ledger which suddenly intrigued him.

She laughed her indignation, a bellowing *wahh-hh* that struck the glass cases like the flat of a hand. "You a *merciless* man for sure."[100]

Black neighborhoods did not have enough political clout to offset their economic weakness—a fact that does not mean black politicians were powerless. Some got black lawyers appointed to various city commissions and to the DA's office. Others had enough pull to protect gambling rackets, get licenses for cab drivers, and procure jobs for local people in the building of public housing. A select few, like Chicago's William Dawson and St. Louis's Jordan Chambers, wielded real power. Chambers, for example, had a hand in helping thirty-nine blacks gain elective or party office by the late '50s, including the first black voted to the board of education in a city-wide election, the first black appointed to the state's circuit court, and the first black in Missouri to win election to the state senate. But on the whole, black politicians secured much less patronage than their white counterparts and so distributed few jobs in the city bureaucracy, on public works projects, and in the police and fire departments.[101]

Black leaders also had to decide whether to promote or prevent public housing, whether to fight for more patronage or push for civil-service reform, and whether to build new schools and hospitals in the black community or try to integrate existing facilities outside of it. Upper-class leaders in Newark's black community, for example, concerned themselves primarily with civil rights. That put them at odds with black councilman Irvine Turner, who wanted the black vote in the central ward and so tried to deliver as much public housing and as many jobs as possible. He grew impatient with black "eggheads," who, he thought, liked to "study the problem" and ingratiate themselves with powerful whites. Jordan Chambers fought with St. Louis's NAACP leaders. Chambers wanted jobs for constituents, and he preferred straight patronage deals to fighting for civil-service reform. He was very independent and would deliver votes to whoever would do favors for him. Black politicians tied to Chicago's machine supported public housing. They tried to help their constituents relocate or get better deals if they were forced to sell their properties. Other blacks fought against urban renewal. The issue created serious rifts inside the NAACP and the Urban League.[102]

To blend different kinds of people into a community is difficult under the best of circumstances, and it was especially hard in the black districts. The middle and upper classes had too little economic or political power to tie together much of the rest of the population through jobs or political patronage. The lower two-thirds of the population got the city's worst jobs and its poorest housing. Certain families had been in the city for generations, some had a couple of decades under their belts, and many others had little experience in urban life. Labor unions, meanwhile, discriminated fiercely against working-class blacks, while most black professionals had to work in segregated institutions or businesses. Whites never had to build their neighborhoods out of so many different kinds of people, nor under the duress of such fierce prejudice.

MAKING PROGRESS

Many black residents felt deeply attached to their neighborhoods. Asked in 1950 why he lived in Harlem, an ice-cream vendor said, "I know everybody in my block and I don't think I want to go anywhere else to live until I get to heaven." Another man replied, "What a question! I can't find anywhere else to live, and if I could I don't think I'd want to live there. I'm originally from Daytona, Florida, but I've been in Harlem for seventeen years. I wouldn't know how I'd feel living any place else."[103]

Blacks in northern cities enjoyed certain freedoms unknown in the urban South and yet still suffered prejudice. Baltimore, which straddled the line between North and South, had the qualities of both worlds. The *Baltimore Sun,* for example, "hired no blacks except for housekeeping jobs and covered practically no black news. Murders of black people were not 'little murders,'" recalled journalist Russell Baker. "They weren't murders at all, as I discovered . . . on phoning the city desk with details of a man who had died of head injuries after being bludgeoned. 'You can't hurt 'em by hitting 'em on the head,' said the night editor, hanging up on me." Yet Baltimore offered, on the whole, more freedoms for blacks than did Memphis, New Orleans, Washington, or even Atlanta, which had a large black middle class and several black colleges. "Negroes of Baltimore," reported *Fortune,* "now can work as bus drivers, firemen, baseball players, librarians—all quite as impossible in 1945 Baltimore as (with the rarest exception) in 1956 Atlanta." Blacks could sit anywhere they wished at Ford's, Baltimore's only theater; they could eat at downtown lunch counters; and they could swim or golf in the public parks. In Atlanta, on the other hand, the "Negro is still tightly segregated. . . . Socially, the Atlanta

Negro must take his vacation in a northern city in order to enjoy the luxury of being able to take his children to a downtown movie." Atlanta's banks did not hire blacks, and manufacturers like Ford and GM employed but a token few.[104]

Washington was no better. President Truman thought of appointing Ralph Bunche as assistant secretary of state. Bunche, who then lived in New York and worked at the United Nations, declined. "I served my exile there," he said of the capital. "Now I prefer to live as a free man." He spoke of "exile" because all of Washington's black children attended segregated schools in 1954. Only a quarter of the city's public playgrounds were open to both black and white children. And many hotels in the nation's capital would not accommodate blacks.[105] Segregation was the norm throughout the South. "Negroes have helped in the development" of Memphis, fumed Reverend George Long of the Beale Street Baptist Church; "but Negroes have no part in the ruling of Memphis. Negroes are not part of the city administration; not upon the board of education; he is not a policeman; he is not a fireman; he is discriminated against."[106] A resident of New York City enjoyed many things about his trip to New Orleans. But he could not abide stepping up to an idle cab, reaching for the door handle, and being told, " 'Ah cain't drive your kind!' " He had to call for a "colored cab." That was in 1962.[107]

Washington, Atlanta, Memphis, New Orleans, and even Baltimore were deemed inhospitable by blacks who had lived in cities like New York and Philadelphia, Pittsburgh and Cleveland, Chicago and Milwaukee, San Francisco and Los Angeles, even Kansas City and St. Louis. A fictional character in a book by Langston Hughes says,

I would not move for no depression. No, I would not go back down South, not even to Baltimore. I am in Harlem to stay! You say the houses ain't mine. Well, the sidewalk is—and don't nobody push me off. The cops don't even say, "move on," hardly no more. They learned something from the Harlem riots [of 1943]. They used to beat your head right in public, but now they only beat it after they get you down to the stationhouse. And they don't beat it if they think you know a colored congressman.[108]

The freedom not to have your head beat in by a cop on the street may not seem like much. But as a freedom lacking in most of the South, it meant a great deal to those who had moved up north.

The North's economic freedoms meant a lot, too. In one of the "forums" held among its nationwide readers, the *Pittsburgh Courier* asked, "Are economic advantages better for Negroes in the North than in the

South?" Three-quarters of those living in northern cities said yes; about 10 percent were undecided.[109] They had good reasons for saying yes. Upward mobility by blacks in Pittsburgh's job market, for example, exceeded downward movement by more than two to one in the 1940s and 1950s. The percentage of blacks who had at least semiskilled jobs in Boston increased from one in eight in 1940 to one in three by 1960. More black professionals worked in every northern city after the war. Most of them were teachers, but some were doctors, lawyers, and engineers. Black women in San Francisco moved out of domestic work and into schools, banks, department stores, and insurance companies. Even small openings in the labor market meant a lot to black wives, who were 50 percent more likely to work than white wives.[110]

Between 1940 and 1960, the job market widened for blacks, their per capita income doubled, and the gap between the average incomes of whites and blacks may have narrowed slightly.[111] The authors of *Black Metropolis* noted the improvement:

The Negro population of Chicago doubled between 1950 and 1960. Bronzeville's doctors, dentists, lawyers, preachers and businessmen captured a sizeable share of the dollars circulating in the expanded and, to some extent, "captive" Black Belt market. Their offices, churches, and homes reflect the new prosperity as do newspaper notices of frequent continental and occasional intercontinental travels, and the elaboration of social rituals.[112]

Despite such advances, American cities were still, as *Newsweek* wrote of New York in 1959, "allergic to color." Black workers found it hard, for example, to join a union. While some union leaders, like George Meany, talked about integration, most union members thought of themselves as white men first and union men second. The building trades were extremely hostile toward blacks during the '50s, and the constitutions of the four big railroad unions had clauses that barred workers on racial grounds. Blacks had a hard time getting white-collar jobs, too. The number of black urban professionals did increase during the '50s, but the rate of increase was less than it was among whites, and black professionals earned less than their white counterparts. Even secretarial schools told black applicants they would be wasting their time because it was nearly impossible to place skilled black secretaries. So even if the black metropolis was said to have "become a gilded ghetto" in the '50s, to many of its residents it was still "a ghetto all the same."[113]

While the labor market opened somewhat in the '50s, the housing market remained highly segregated. Many blacks found better places to live, but most of them did so by moving into areas vacated by whites. Others got better accommodations by moving into public housing. It is true that the projects displaced more people than they housed. They certainly destroyed many buildings that could have been renovated. Many of them were segregated by law until the middle of the '50s. Any sense of community, moreover, was obliterated along with the old houses, apartment buildings, and corner stores.[114] Yet most people favored public housing. The units were new, clean, and efficient, and it seemed as if government was finally doing something for black neighborhoods. Mary Wilson of the singing group the Supremes may have had the better part of a common experience: "Though the average person might automatically think of anyone living in public housing as being deprived, we barely knew the meaning of the word. Our parents constantly reminded us that we were far better off than they had been at our age, and the camaraderie and freedom we felt as kids contributed to making mine perhaps the first generation of black youths to believe their individual potential was unlimited."[115]

Although many blacks found better places to live during the '50s, a majority listed housing as their primary concern. They did so for good reasons. The Supreme Court ruled against restrictive covenants in 1948, but they remained in effect in many neighborhoods. The Federal Housing Administration did not back mortgages in mixed or transitional neighborhoods, because it stressed racial unity as a criterion for guaranteeing loans. Many blacks bought houses through lending companies that repossessed units after one missed payment. For every person who got nicer housing on the edge of the expanding black community, or in the new projects, many more lived in substandard and overcrowded units neglected by landlords. No matter where they lived, blacks paid on average about 10 percent more than whites for similar shelter. Neighborhoods that turned from white to black became much denser because two or more families typically moved into what had been a single-family house and most apartment buildings were cut up into smaller units. Every black person, finally, knew that just moving to a white or a transitional neighborhood meant trouble of some sort.[116]

The slow but notable progress in job markets and living conditions was part of something bigger in the 1950s: a genuine feeling among blacks that America would finally live up to its constitution by making all of its people, at least in legal terms, full and equal citizens. "In their continuing struggle for complete acceptance in the American scheme," wrote the *Chicago Defender,* summing up the previous year's events, "Negroes may well come to consider 1956 as the decisive year, a period when the fight turned definitely—and irrevocably—in their favor." Some white Americans cherished the goal of the complete acceptance of blacks in the American scheme; many others feared it. "For more than 17 million American Negroes," warned *Fortune* magazine in 1956, "this freedom may not be total and flawless: a number of white citizens, from southern governors to northern real-estate brokers, can be counted upon to see to that."[117]

The federal government and the Supreme Court helped to foster a cautious optimism. Harry Truman, for instance, was the first president to invite blacks to all functions of an inauguration, to appoint a black to the federal judiciary, to address a meeting of the NAACP, and to campaign in Harlem.[118] In 1954, the Supreme Court declared unconstitutional the doctrine of "separate but equal." Congress signed into law a weak civil-rights bill a few years later.

Those national events affected American cities, they were reported by the black weeklies, and they coincided with certain improvements in the neighborhoods. Every northern city, for example, had at least a few mixed schools with a dedicated group of black and white teachers working together. Many northern cities could also report, as San Francisco did, its first black fireman, its first black housing-authority commissioner, its first black assistant district attorney, and its first black city clerk.[119]

Black neighborhoods in most large cities gained political clout as well. What was written about Chicago was true of many other cities: "The addition of new neighborhoods to the Black Belt has increased the number of new aldermen in City Hall, and has resulted in more precinct captains, ward officials, and 'jobs for the boys,' as well as symbolic appointments to high office." Black congressman William Dawson ran a powerful political machine on Chicago's South Side. He expressed the mood of the mid-'50s when he said, "The Negro is a citizen. I have faith and confidence that every right of citizenship will be his some day. We should see all restrictions almost wiped out before many more years."[120]

The middle and upper classes—dentists, doctors, lawyers, teachers, undertakers, social workers, store owners, newspapermen, politicians, educated preachers, and radio-station operators—worked especially hard after the war to lift those restrictions and improve the lives of all blacks in their cities. A study of black leaders in San Francisco showed that they, like their counterparts in other cities, "no longer believed they were powerless to improve the quality of their lives, and they willingly shouldered a greater share of the burden of cleaning up their community and making progress."[121]

Progress meant obtaining for blacks the same rights and opportunities already enjoyed by whites. Boycotting buses in Montgomery, contesting a segregated roller rink in Chicago, refusing to move to a Jim Crow rail coach in New Orleans, picketing a Kansas City department store that refused to hire blacks, defending the rights of blacks to move into white neighborhoods, demanding black election officials in black districts, calling for the mayor's office to provide local schools with their fair share— in every case the plea was for white Americans to live up to the Constitution. Not even a majority of the middle and upper classes wrote in the black weeklies, worked for the NAACP, protested in the streets, fought in the courts, donated time to charitable organizations, or made their case in front of school boards. But enough did to make themselves and their causes visible in their own neighborhoods and in their cities at large.[122]

The civil-rights movement took place primarily in cities between the late '40s and the early '60s. The principle of equality before the law was enshrined in the Civil Rights Acts of 1964 and 1965. That principle, so forcefully yet gracefully embodied in Martin Luther King, reflects the best of America. King's idea was to assimilate blacks fully into American society—to get them into baseball, politics, movies, housing, unions, and more. But in cities like Nashville, as late as 1960, it was still a problem to get into a restaurant:

At the counter were the Negroes, not talking to each other, just sitting quietly and looking straight ahead. Behind them were the punks. For more than an hour the hate kept building up, the hoodlums becoming increasingly bold. The crowd watched appreciatively: "Here comes old green hat," referring to one of its favorite hoods. "Looks like it'll go this time." The Negroes never moved. First it was the usual name calling, then spitting, then cuffing; now bolder, punching, banging their heads against the counter, hitting them, stuffing cigarette butts down the backs of their collars. The slow build-up of hate was somehow worse than the actual violence. The violence came quickly enough, however—two or three white boys finally pulled three Negro boys from the counter and started beating them. The three Negro boys did not fight back, but stumbled and ran out of the store: the whites, their faces red with anger, screamed at them to stop and fight, to please goddam stop and fight. None of the other Negroes at the counter looked around. It was over in a minute.[123]

The Civil Rights Acts were the culmination of twenty years of court cases, street protests, impassioned pleas, and courageous encounters like the one in Nashville. But as that movement for social justice hit its peak, the neighborhoods passed their turning points. Cities lost the industrial jobs that could have been means of continued progress for many black residents. Middle- and upper-class blacks fled the old neighborhoods, taking with them much of the money that had supported local businesses and many of the institutions that had consolidated the community. Poverty became more concentrated, bulldozers razed block after block in the name of redevelopment, and once lively commercial and entertainment strips became zones of blight. When the neighborhoods declined, so did their music. "To revisit New York in 1963," wrote a jazz fan, "was a depressing experience for the jazz-lover who had last experienced it in 1960." It happened fast, and it happened in cities all across the nation. Whereas the weekly newspapers sought to appeal to all blacks in the '40s and '50s, new weeklies in the '60s reflected the breakup of the black metropolis into more distinctive local communities. A significant portion of those coming of age in the '60s rejected the insistent but patient leadership, and the faith in legal means and peaceful integration, that had characterized the previous generation. A once hopeful anger turned into desperate rage. That rage seared black neighborhoods all across America in the second half of the '60s.[124]

FIGURE 23. Waiting for a streetcar in Pittsburgh, 1951. Courtesy of the Carnegie Library of Pittsburgh.

FIGURE 24. A San Francisco commercial avenue, full of small stores and neon signs, 1964. Courtesy of San Francisco History Center, San Francisco Public Library.

FIGURE 25. The caption of this 1951 photograph of Pittsburgh read: "Tradesman talking with butcher in front of his shop." Courtesy of the Carnegie Library of Pittsburgh.

FIGURE 26. Greengrocer with customer in New York City, ca. 1955. Photograph by Ronald Dubin. Courtesy of the Museum of the City of New York.

FIGURE 27. A new supermarket in San Francisco, 1950. This market was an amalgam of several small businesses, including a baker, a grocer, and a butcher. Courtesy of San Francisco History Center, San Francisco Public Library.

FIGURE 28. Another new supermarket in San Francisco, 1959. The newscopy read: "SHOPPERS NOTE: There are plenty of parking spaces—149 above—at Safeway's ultra modern store." Courtesy of San Francisco History Center, San Francisco Public Library.

FIGURE 29. A neighborhood restaurant in San Francisco, early 1950s. Courtesy of San Francisco History Center, San Francisco Public Library.

FIGURE 30. A stoop, a man, and a woman on the cover of Irving Schulman's novel *Cry Tough!*, 1950. Courtesy of Avon Books.

FIGURE 31. A vagrant in a doorway, New York, 1952. Photograph by Larry Silver. Courtesy of the Museum of the City of New York.

FIGURE 32. Workers ripping up streetcar rails, San Francisco, early 1950s. Courtesy of San Francisco History Center, San Francisco Public Library.

FIGURE 33. Factory with rail connection in a manufacturing-residential district of San Francisco, 1948. Courtesy of San Francisco History Center, San Francisco Public Library.

FIGURE 34. Redevelopment on Manhattan's Upper West Side, 1959. Photograph by Frank Paulin. Courtesy of the Museum of the City of New York.

FIGURE 35. The back of this photograph reads: "Bull session at Jimmie's, 1130 Fillmore." San Francisco, 1964. Courtesy of San Francisco History Center, San Francisco Public Library.

FIGURE 36. A shopping street in Harlem, ca. 1950. Photograph by William Glover. Courtesy of the Museum of the City of New York.

FIGURE 37. Shopping street on the western edge of Oakland's business district, adjacent to a black neighborhood, ca. 1955. Courtesy of the Oakland Public Library, History Room.

FIGURE 38. A gathering of Pittsburgh's black upper classes, 1951. The original caption read: "The 40th Annual Dinner of the Centre Avenue 'Y.M.C.A.' held in the gymnasium of that organization. The dinner was attended by active laymen and leading citizens of Pittsburgh." Photograph by Richard Saunders. Courtesy of the Carnegie Library of Pittsburgh.

FIGURE 39. A Sunday afternoon in Pittsburgh, 1951. Photograph by Richard Saunders. Courtesy of the Carnegie Library of Pittsburgh.

FIGURE 40. A young family leaving their apartment in a building about to be destroyed for public housing, 1951. Photograph by Richard Saunders. Courtesy of the Carnegie Library of Pittsburgh.

FIGURE 41. An alley in a black neighborhood in Pittsburgh's Hill District, ca. 1955. Courtesy of the Carnegie Library of Pittsburgh.

CHAPTER 4

The Suburbs

The face of suburban house builder William Levitt was on the cover of *Time* in 1950. The caption read, "For sale: a new way of life." That sounds like hyperbole, but it was true. All of the houses in the suburbs were new. So were most of the families, appliances, and living-room sets. Mothers drove to brand-new shopping centers. Kids attended new schools. Residents even created a new kind of community. For suburban dwellers did not shop on a pedestrian commercial avenue, they did not live in a place with a sense of history, and they did not share with neighbors a common ethnic culture. Instead, the new community was built around a broad middle-class identity, the activities of children, heavy participation in voluntary groups and associations, and the automobile and the shopping center. It was easy to see, as novelist John McPartland did, that suburbanites were "new, new, new. Like no other people who had ever lived." It was harder to see, as *Fortune* did in 1953, that suburban residents were "developing a way of life that seems eventually bound to become dominant in America."[1] The magazine was right. The culture and geography of the new suburbs have come to dominate society today.

The Inhabitants

The housing stock of nearly every city was full in the late 1940s and early 1950s. Little had been built between the end of the 1920s and the end of the war. Some fifteen million members of the armed forces returned to civilian life in 1945 and 1946. Southern blacks migrated to northern cities during and after the war. Everyone married young and had kids early. No wonder six million families stayed with friends or relatives in 1947, while another half a million occupied makeshift dwellings such as converted Quonset huts. In New York City alone, more than 165,000 families doubled up after the war, and thousands of newlyweds stayed with in-laws.[2] Many others lived in expensive but tiny apartments: "When we were married in 1948," a man recalled, "my wife and I moved into a third-floor walkup, one-room, kitchen-in-the-closet, 'efficiency' apartment in a subdivided brownstone on West Fifty-seventh Street [that] rented for seventy dollars a month. We knew that we were being robbed, but still we were glad to have found it." They were gladder still, according to city planner Robert Moses, when they bought a house in the suburbs: "To young couples with one or two small children, fed up with life in a city sardine can or goldfish bowl, often with elderly in-laws, any house or apartment of their own in any subdivision looks like Heaven." A house in a suburb looked especially divine when the new owners wrote a monthly mortgage check for no more than they had been paying their landlord.[3]

This passage from John McPartland's 1957 novel, *No Down Payment*, illustrates one couple's move to a suburb:

> Eight months ago, Betty and he and the kids in the rented house on Nineteenth Street. . . .
> "Herman—have you read anything about that Sunrise Hills development? Down off the Bayshore?" the area supervisor had asked him one morning.
> "Not particularly," he had said. . . .
> "Several thousand homes and a shopping center. We're putting a branch in the shopping center—the whole appliance line, tires, car accessories. We're figuring you for manager, Herm."
> That was the way it had been.
> "Of course, Herm, there are some details. We like our managers to take part in civic activities—join Kiwanis or Lions, maybe even Rotary. You'd just about have to live in Sunrise Hills, buy a place."[4]

Herman and his wife were typical of movers to suburbs: they were young at a time when the average age of adults in suburbs was just over thirty for husbands and a few years under thirty for wives.[5] Herm, his wife, and their two young children moved from a rented house in the city. Couples typically moved with a small child or two from either an apartment, a rented house, or shared quarters with in-laws or siblings. Herm and his wife grew up in the city and had never lived in a suburb. Most couples experienced the suburbs as first-time homeowners. Herman and his wife were white, second-generation Americans. Almost everyone in suburbs had been born in America to parents of European descent. Herm graduated from high school and was an assistant manager in the appliance department of a big store. A few suburbs were made up of college-educated couples whose lives revolved around the local country club. Other subdivisions housed mostly blue-collar men who worked in nearby factories.[6] But most suburbs had the full range of a broadening middle class, men like Herm who worked as salesmen, teachers, city clerks, cops, firemen, unionized factory workers, supermarket and department store supervisors, and lower- and middle-level staff in large companies. In the Levittown subdivision outside Philadelphia, for example, a fifth of the men earned their livings as professionals, half had white-collar jobs, and a quarter were blue-collar workers. Less than 10 percent had a college degree.[7]

Herm, his wife, and the kids moved into a house that was pretty much like all of the others. Some people chose Cape Cods or split-levels, but most lived in ranch houses, with their low-pitched roofs, deep eaves, picture windows, and strong horizontal lines. Slight variations in layout and ornament—which tended to repeat every six or seven houses—distinguished one house from the next. Whereas many city neighborhoods had a great mix of architectural styles or a very distinctive kind of housing—like the triple-deckers of New England, the Victorians of San Francisco, the row houses of Baltimore and Philadelphia, the brick bungalows of Chicago, or the brownstones of New York—a new suburb in one part of the country looked like those in any other. Many suburbs were mass-produced with materials made in city factories. The job was divided into a few dozen tasks. White-paint men, red-paint men, tile layers, wall builders, roof makers, and bathroom installers executed their special tasks. Men cut and assembled materials in central workshops and dropped them off at the site. Drywall replaced plaster. Skilled laborers made up just 20 percent of all workers. Some developers even installed stoves, refrigerators, and television sets that they bought directly from manufacturers.[8]

No wonder *Fortune* magazine, in 1953, regarded suburbs as ideal places for an emerging class of transient employees moved around regularly by their companies: the landscapes looked pretty much the same. The suburbs had the additional advantage of little, if any, settled culture or history, and so there were no established social hierarchies. Their homogeneity, in other words, made them more or less interchangeable and thus easy to adjust to.[9] That is exactly what critics like Lewis Mumford complained about:

In the mass movement into suburban areas a new kind of community was produced, which caricatured both the historic city and the archetypal suburban refuge: a multitude of uniform, unidentifiable houses, lined up inflexibly, at uniform distances, on uniform roads, in a treeless communal waste, inhabited by people of the same class, the same income, the same age group, witnessing the same television performances, eating the same tasteless pre-fabricated foods, from the same freezers, conforming in every outward and inward respect to a common mold, manufactured in the central metropolis.[10]

Mumford's overwrought censure contained some truth. Everyone furnished their homes with the same new products. One suburb looked pretty much like the next. Most adults were young. The absence of blacks reflected a time not yet ready for integration.[11] Distinct ethnic and cultural aspects of churches were observed to be blending into a single, indistinct suburban church.[12] Different levels of income did separate some subdivisions from others. And cultural markers like blue-collar versus white-collar, or college-educated versus high-school graduate, did distinguish some families from their neighbors. But class was a weak social emblem in these new suburbs, which joined together the classic American forces of cultural assimilation, economic mobility, and ownership of property. Those forces, wrote William Whyte in 1953, "ironed out the regional and religious differences and prejudices that have separated people."[13]

Because this converging way of life made class, ethnic, and religious indicators somewhat vague, it was "the little differences," as one journalist called them, that decided who were friends, who belonged to which groups, and who was ostracized.[14] Generally speaking, those differences were matters of temperament rather than class, of looks rather than family background, of personal taste rather than religion or ethnicity. They were matters, in other words, of bowling or softball, of Budweiser or Dewar's, of playing cards or talking politics, of whose kids played together, and of whose wife coveted whose husband. What else but "little differences," after all, could have distinguished one person from another in fairly homogenous suburbs that lacked the city neighborhood's old-timers and extended families, its longtime rivalries and long-standing friends, its ethnic clubs and churches, and its local drunk, whore, and idiot?

The Household

The 1950s are commonly seen as the last days of the traditional family. Parents did have more authority than they do now, and as a rule men earned wages while women worked at home. But the suburban family of the '50s was as new as it was traditional. It focused on itself more than families did previously, and it connected itself even more closely to social, civic, and religious groups than did families in the old neighborhoods.[15]

The main reason for an intense focus on family life in suburbs—besides the great numbers of young parents with children—was the household's isolation from a neighborhood culture or a small-town community. Instead of familiar and routine relations with small merchants on a nearby retail street, for example, most suburban families drove to shopping centers for impersonal exchanges in large stores. Whereas the old neighborhoods contained extended family members, suburban kids did not play with cousins down the street or walk to their grandparents' place. Nor did new suburbs provide the sense of permanence that city neighborhoods and small towns gave to their inhabitants. As a consequence, many families tended to focus on themselves. Parents got involved in their children's activities and doted on them as never before. Young husbands and wives had more reciprocal relationships than did previous generations of parents. The typical suburban home had more room for the kids than the average city apartment or house; it was filled with all sorts of goods, including a television set; and it strove for what *McCall's* described as "togetherness."[16]

None of that is to say the average family cut itself off from the outside world. Enough residents participated in groups and associations for one observer to call the new suburbs a "communal way of life."[17] Here was the key to the cultural life of suburbia. Upon moving to a suburb, residents lost daily contact with extended family members, their neighborhood culture, and a retail street they could walk to. Young couples made up for those losses by focusing more intensely on the family and by linking it to all manner of religious, charitable, political, and recreational groups and activities. So while suburban households had no ties to a block, a retail corner, or an ethnic culture, they did tie themselves to a community through membership in clubs and associations.

Before someone set up house, however, they had to get married. A higher percentage of people were married in the 1950s than at any time since the late 1800s. The median age of marriage (just under twenty-three for men, just over twenty for women) was lower than at any time since at least 1890. Marriage was a much more compelling norm in the '50s than it had been ten, twenty, or thirty years earlier, or has been since.[18] "I wasn't old enough for anything," remembers Mary Cantwell. "But marrying young, a classmate used to say, was like getting to a sale on the first day. God knows what, if anything, would be left if you waited till you were twenty-five or -six."[19] Close-ups of melting girls and ardent young men in tender embraces illustrated page after page of serialized magazine stories. Although they did not marry until the end of the story, matrimony was always the conclusion. To one woman looking back on those years, the "current of the mainstream was so strong that you only had to step off the bank and float downstream into marriage and motherhood." Even bachelors felt the pull. The society at large regarded a single man over thirty as either a rake, an emotionally disturbed person, or a pitiful creature fettered to his mother.[20]

The ideal girl was a virgin on her wedding day, and that ideal was not far from reality. Alfred Kinsey's 1953 report *Sexual Behavior in the Human Female* showed that while up to 95 percent of American women had "petted" before marriage, only about half had lost their virginity before their wedding day. Among those who had lost their virginity, most had done so with their intended husband. Three times as many men as women, on the other hand, reported having had intercourse with someone other than their future spouse. If young women as a whole lacked sexual experience, they also lacked sexual knowledge: "[G]oodness knows we never discussed sex," remembers Mary Cantwell. "Doing so would have implied that the speakers knew something about it, and if we did, we 199 graduates of Connecticut College Class of '53, we kept it a secret." Writer Nora Johnson, however, said that the average girl had done "every possible kind of petting without actually having had intercourse."[21]

It is not certain why people married so young, bought houses, and filled them, on average, with three or four children, but it is clear that it was economically feasible. There were plenty of jobs, incomes rose steadily, the price of a house was low, and the federal government guaranteed mortgages. Some writers saw early marriage and home ownership as a search for security by a generation that had grown up during a depression and had experienced war as young adults. Others saw the hunkering down as a reaction to the increasing size and power of corporations, unions, and government, which put greater distance between the average person and the realms of work and politics. Margaret Mead blamed it on the conservative political culture of the '50s: "Lacking any kind of leadership young people had another date, found a new job, got married, and concentrated on having babies, one after another."[22]

Whatever the reasons for early marriages and large families, the household operated around a clear division of labor: men worked for wages and women worked at home. In 1950, about 23 percent of all married women living in metropolitan areas—just over 30 percent of black and just under 20 percent of white women—worked outside of the home. A clear trend was for newly married women to work until they had their first child: only about 10 percent of white mothers with preschool-age children worked. The likelihood that a mother with preschoolers worked outside of the house depended on her husband's income. Whereas just 3 percent of mothers with preschoolers in families making more than five thousand dollars (a middle-class income in the mid-'50s) held a job, 16 percent of those in households earning less than that worked for wages. While there are no precise figures on the work histories of suburban women, almost all women in suburbs were married with young children, the average suburbanite earned more than the average city dweller, and very few blacks lived in suburbs. So it seems reasonable to assume that, on average, only about one of every ten suburban households had a woman in the work force at any given time during the '50s.[23]

Four typical households, arranged by income, would look something like this. In 1955, *Life* chose one woman as the "composite" of the Vassar class of 1940. She was married to an Ivy League graduate, she did volunteer work with the League of Women Voters and the PTA, and her three children attended school. She did not work. But about one in seven of that 1940 class, whose kids were all in school by 1955, did have at least a part-time job. Down a notch economically, and younger by ten years, was an educated couple in the upper tiers of the middle class. The man worked as a low-level executive in a Republic Steel plant and the woman worked full-time at a bridal service in a department store. A housekeeper watched their young child. The mother worked for personal satisfaction. That arrangement probably ended soon, however, for she likely had one or two more kids at a time when finding maids and baby-sitters was hard. Next came the broad middle class, which made up the great majority of those living in suburbs. The male was a unionized blue-collar worker, a school teacher, a fireman, a manager of a supermarket, or a white-collar man in a corporation. The typical wife stayed at home. Those wives who worked did so after the youngest child had entered school, and at a job with hours flexible enough to allow her to see the kids off in the morning and return home when school let out. At the bottom of the suburban hierarchy were working-class families in which both parents worked. "Peggy and Ralph work long hours and are away from home all day," wrote *Fortune* magazine about one such couple; "but they are not too tired to enjoy the evening with the family and to plan for a better future." They had a modest house and two little girls. She had a job at an aircraft plant on Long Island, while her husband, who had worked in a lumber company, was trying to set up a roofing business with a partner. Their mothers took turns watching the kids, which was not easy because both mothers lived in the city.[24]

The typical suburban mom stayed at home, then, just like her mother or grandmother. But her job as a homemaker became something special. Kids, after all, deserved special treatment. Most suburban couples had experienced the Depression as children or adolescents, many were young adults during the war, and some had gone overseas. As new parents in a great economic time, they had the means to give their kids things that they themselves did not have as children—such as a backyard, plenty of toys and activities to keep them amused, and mothers to drive them around. Magazines wrote constantly about the great amounts of time and energy that young parents devoted to their children.

While most people agreed that the extra comfort and attention were good for kids, many people suspected that this new generation of children might be getting too much of a good thing. One writer, for example, feared the arrival of what many grandparents already saw: "a race of demanding, nagging, overindulged, overpampered kids—and a race of beaten-down, spineless parents who take their orders." Another complained that "we now look on them not as just 'kids,' as we used to, but as a sub-culture with a powerful effect on the culture as a whole." An otherwise restrained observer predicted that history would show "we are a buffer generation, standing by silently while our children, brought up by demand-feeding and demand-everything, kick over the traces and do startling things, with none of our predilection for playing it safe."[25]

The attention lavished on kids was a result, in part, of changes in behavior between wives and husbands and between fathers and children. It is hard to see those changes today, because we see the fathers of that time as so uninvolved and the division of duties between men and women as so stark. Journalist Russell Baker, for instance, remembers coming "of age in this astounding prosperity when a husband's role was to go to the office and become a success while the wife's job was to get the children born and smartened up for a really good college. Men from this world obviously wouldn't give fatherhood the priority, the energy, the study, and the seriousness it required."[26]

Baker was right—when viewing the '50s from the vantage point of the late '80s. But most young husbands in the '50s gave fatherhood more time than their own fathers had, and many women demanded a larger say in their marriages. "No change in the American family," wrote the authors of a book about marriage in the '50s, "is mentioned more often than the shift from one-sided male authority to the sharing of power by husband and wife. Perhaps no change is more significant, either. . . . [This is] a new generation of American wives who are more resourceful and competent than their grandmothers. They are no longer content to sit quietly by while their husbands make the decisions."[27]

Life reported on "the new American domesticated male" in 1954. The piece showed husbands and wives working as partners. The man, it said, "has become baby tender, dishwasher, cook, repairman. Probably not since pioneer days, when men built their own log cabins, have they been so personally involved in their homes." The children's education was "no longer sloughed off on the wife," and many fathers joined the PTA. In another story, *Life* documented the activities of a father who spent a weekend caring for his infant twins, as well as his three- and nine-year-olds, while his wife visited her parents. The article showed how hard it was to manage a house. "It never ends," the father complained to a friend; "that's what's discouraging." In addition to helping out around the house and volunteering time to Little League and Boy Scouts, young fathers also got behind the wheel of that emerging summer tradition called the family vacation. The number of visitors to national parks, for example, increased from sixteen million in 1940 to a hundred million in 1964.[28]

Of all domestic jobs, however, men liked best to cook in the backyard. *Life* helped them out, in 1953, with an article called "How and What to Grill." Novels, television ads, and magazine articles depicted suburban man wielding his long fork beside a setup of salt and pepper, lighter fluid, and auxiliary utensils. His wife may have patted the hamburgers and marinated the steak, but there he was, in his backyard, cooking the meat, poking the potatoes, and toasting the buns. They bought all of the ingredients at the shopping center: the grill, the charcoal, and the lighter fluid; the hamburgers, hot dogs, and rolls; the mustard, relish, and ketchup; even tin foil and paper plates. A character in a John Cheever short story thinks about standing over the grill, fork in hand, the kids playing nearby, his wife providing materials and seasoning: "We have a nice house with a garden and place outside for cooking meat, and on summer nights, sitting there with the kids and looking into the front of Christina's dress as she bends over to salt the steaks, or just gazing at the lights in Heaven, I am as thrilled as I am thrilled by more hardy and dangerous pursuits."[29]

Less thrilling, but possibly more satisfying to fathers, was the time they spent on their lawns and houses. In many sections of the country, most men spent at least a part of every summer weekend mowing, applying chemicals, and digging out (or having their kids dig out) dandelions with jackknives. Men enjoyed yard work and took pleasure in a handsome lawn, which was both a way of showing respect for neighbors and a means of maintaining property values. Millions of men also used the new power tools (they spent 6 million dollars on them in 1947, 150 million in 1953) to add backyard decks, make bird feeders and custom mailboxes, and build rooms in basements. Such touches mattered in suburbia because one's taste and possessions were more visible than in the city. The condition and decoration of a house was conspicuous, landscaping skills were on display, and visiting was common.[30]

The work that a man did around the house, the time he spent with the kids, and the more equitable relationship he had with his wife shaped the character of the suburban home. But these domestic changes only went so far. Even if, as one woman remembers, most young husbands "were involved with colic and toilet-training and diaper rash as their own fathers never were," and even if they "did change diapers and scrub greasy pans at night and on weekends," it was also true that "their wives would remind them that they were away all day, far from the wails and the whine of the vacuum cleaner." Men could not quite understand, another woman complained, that housework is never done.[31]

Nor did these domestic shifts work themselves out evenly across society. "During an era of changing attitudes," concluded a study of working-class marriages in the 1950s, "the question of the husband's responsibilities might be expected to cause some conflict in marriage. Should a husband be expected to help with the dishes, to scrub the kitchen floor occasionally or to take turns during the night with the baby's bottle? For eight out of every ten of these working class couples 'who does what around the house' does not constitute a troublesome issue." Most working-class couples, in other words, did not think the man ought to help out around the house. Raymond Chandler's character Philip Marlowe may have been speaking for millions when he scorned the tired men driving "towards home and dinner, an evening with the sports page, the blatting of the radio, the whining of their spoiled children and the gabble of their silly wives."[32]

Yet the shift in domestic relations went far enough to inspire new public images of husbands and wives. When compared to comic strips in 1929, for example, those in 1956 more frequently portrayed men as dominated by their wives, henpecked, and immature. Similarly, cartoons featured the American male as a second-class citizen in a matriarchy: females ran the house and spent the money while the men took refuge in their little worlds of baseball, lawns, and cars. While parody is not reality, these caricatures played on the fact that younger husbands, as a whole, built their lives much more around their wives, children, and houses than their own fathers had. *McCall's* described the average suburban father in 1954: "Had Ed been a father twenty-five years ago, he would have had little time to play and work along with his children. Husbands and fathers were respected then, but they weren't friends and companions to their families. Today the chores as well as the companionship make Ed part of his family. He and Carol have centered their lives almost completely around their children and their home."[33]

If the typical suburban mother expected her husband to help out around the house more than her own father had, she also enjoyed a house that was stocked with the latest of what *Fortune* called an "upheaval in home goods." The main character of John McPartland's novel *No Down Payment* marveled at the houses around her and the home goods that filled them:

All of them new, with redwood, fieldstone and glass, all of them with patios and shining kitchens, television sets, automatic washers and dryers, automatic mixers, coffee-makers, toasters, garbage disposal gadgets, clock radios, freezer cabinets, automatic lawn sprinkling systems, electric lawnmowers, dishwashing machines. Nothing down and not much a month, and everybody young, secure, with the healthiest children, all of them still little and fun, full of vitamins and whole wheat cereals, drinking endless quarts of extra-rich milk.[34]

Such enthusiasm was natural for someone who had grown up in a depression and a war economy, yet who raised her family in a time when working- and middle-class incomes were growing, government-backed mortgages were allowing tens of millions of people to buy houses, and manufacturers were pumping out affordable products that were standard equipment in most new homes. Refrigerators, washing machines, living-room sets, modern bathroom fixtures, and kitchen appliances—this great output of "home goods," wrote *Fortune,* "has working for it some of the most potent forces in American life: the love of the new, the fascination with anything mechanical, the rising interest and pride in Americans' surroundings." Those home goods were "rapidly becoming the exemplar of the new, homogenous American market; in no other area has the old mass-class division been broken down so thoroughly."[35]

Modern kitchen items, for example, removed distinctions between the masses and the upper classes. These new products exemplified the solid yet sleek style of industrial modernism. The stove was usually powered by electricity and controlled with precision dials. The dishwasher filled a space under the counter. The refrigerator and freezer were neatly organized into compartments and could easily hold a week's worth of packaged food for a family of six. Toasters took on a shape that made them seem to lean forward, ready to surge ahead. Blenders gleamed with steel, chrome, and glass. An electric opener punctured the can, circled the rim, and lifted the lid. Millions of kitchen tables had metal tubes for legs, a broad metal band girding the Formica table top, and tube-leg chairs whose seats and backs were covered in Naugahyde.

Those home goods furnished an open kitchen. The dining area was usually an unbroken extension of the cooking area, making it easy to serve the food. The back or side door, which was often the favorite entrance, opened into the kitchen. Mothers watched their kids play in the backyard through windows above the sink. Such efficiency inspired some to hail the new suburban mother as a household engineer. It also reflected the fact that modern appliances were not just *in* the kitchen; they very much *were* the kitchen.[36]

The average kitchen beamed with colors like Bermuda pink, fern green, sand beige, lagoon blue, and buttercup yellow. Those colors brightened the days of many women, but some mothers were "sated to the point of nausea" with pink-handled brooms, lemon-yellow dishpans, pale-blue Formica table tops, bright red measuring spoons, and bronze-gold canisters.[37] Even the images of food on packaging and in magazine ads were depicted in lurid colors that gave an unnatural look to milk, beef, green beans, and tomato soup.

Such colors seem appropriate for the mostly processed foods that mothers prepared in these modern kitchens. The consumption of ready-made food increased fourfold between 1941 and 1950, doubled again between 1950 and 1953, and continued to grow throughout the rest of the decade. Pink-cake mixes, canned fruits, frozen vegetables, and even complete TV dinners became staples of suburban life. No one took in more processed food than infants. A child under three ate fifty-three pounds of jarred baby food a year in 1953, more than four times what a toddler ate a decade earlier.[38] School kids weren't far behind. They dug their hands into newly opened boxes of Sugar Pops for secret decoder rings. They tore off box-top premiums for plastic bows and vacuum-cupped arrows. They shook the cereal into bright yellow plastic bowls. They covered it with canned pears and then bathed it in vitamin-rich milk poured from glass bottles. Breakfast concluded with a daily vitamin, washed down by either a glass of Ovaltine or a glass of orange juice (which was made by adding water to frozen concentrate that came in a tube). Mother handed the kids their metal, action- or cartoon-character lunch boxes while whisking them out the door. Each box held a bologna, tuna-fish salad, or peanut butter and jelly sandwich made on white bread as smooth as the Formica table top. It held dessert, too, possibly an apple but, more commonly and happily, a Hostess Ho-Ho, Ring Ding, or pink Snowball.

Some suburban mothers made their own pasta sauces, gravies, breads, soups, and cakes. But most evening meals, timed for Dad's arrival home from work, relied on Spam, frozen vegetables, chops from the freezer, and spaghetti sauce from a jar. Mother made brownies and pound cakes by adding milk and an egg to a packaged mix. She let the kids lick clean the metal mixers she had used to blend the frosting, and she topped the dessert off with Reddi-Wip. Consider all of the "frozen foods, the cake mixes, roll mixes, soup mixes and pudding mixes," wrote *Fortune* of suburban cuisine, "and it is plain that a home cook can turn out an elaborate meal with little more effort than it takes to unwrap the packages. If every item on the menu does not taste 'just like mother used to make,' it compares very favorably with the output of the average general houseworker."[39]

It may be doubtful that this new food tasted as good as the average cook's, but it was close enough and it saved time. Frozen dinners, canned vegetables, and cake mixes were easy to prepare. Electric machines opened cans, brewed coffee, mixed ingredients, and cleaned dishes. Large refrigerators with built-in freezers reduced the number of shopping trips and allowed for stocking up during sales. The processed food and the kitchen appliances, along with electric vacuums and washers and dryers, were expected to make time for Mom to become a better mother. Ads admonished her to keep those kids smiling. For that she needed the latest products to rid the house of germs, give shirts and jeans the smell of spring, and keep those kids well fed and ruddy-faced. Mothers at home, in other words, were expected to use the new things of a bountiful economy to make their kids the best little creatures on earth.

No wonder advertisements featured the twin themes of new and modern. "New!" was the most popular pitch. It catered to what a magazine article called the "almost dogmatic optimism" of suburbanites, and their "special delight in the brand-new." The theme of "modern" was close behind. Kitchen tiles were "the modern fashion in floors," living-room carpets were "a triumph of modern luxury," and even sink fixtures were "modern . . . yet time-tested."[40]

New and modern equaled progress. Ronald Reagan assured television audiences that, at "General Electric, progress is our most important product." *Time, Life,* and *National Geographic* validated that progress with photographs of thatched huts in Africa, collective rice paddies and squatting peasants in China, an Argentine air-force jet bombing its own presidential palace, and turtlenecked Europeans driving small, queer-looking cars and living in tiny apartments without refrigerators.

No household item embodied progress quite like the TV. Nearly every suburban house acquired a set in the '50s and made it the centerpiece of either a TV room built in the basement, the family room right off the kitchen, or, in some cases, the living room itself. Suburbs were the first places, according to an article in *Harper's,* "where the impact of TV is so concentrated that it literally affects everyone's life." Cartoons and Westerns, for example, served as excellent baby-sitters. Programming could determine when families ate meals, whether the kids greeted Dad when he came home from work, and how mothers organized their tasks. Engineers actually noted high water pressure during certain shows—like soap operas or Jack LaLanne's exercise program—and low pressure when they ended and everyone went to the bathroom. Social clubs and civic organizations dared not hold meetings during popular shows. Some couples, according to Stan Opotowsky, the author of *TV: The Big Picture,* benefited from the fact that television "increased sexual intercourse, largely because both men and women said they were aroused by the romantic scenes they watched on the late movies." One wife confirmed just that: "Until we got that TV set, I thought my husband had forgotten how to neck."[41]

The most important effect of television, however, was to offer people something else to do besides reading, going to movies, listening to the radio, visiting friends and relatives, or playing in the yard or street. Television may also be a cause of the steady decline, since the '60s, in the numbers of people who belong to churches, social clubs, political organizations, and philanthropic groups. Baby boomers, in other words, not only grew up watching television, but carried over the habit into their adult lives, trading in the civic life of their parents for a more private one.[42]

In addition to being the newest and most modern of all home goods, television instantly became the medium for the greatest boom in the history of advertising. The best way to get a message across to a mass audience is still through a box people watch in their own homes. Messages urged viewers to buy a certain brand of goods. "The greater the similarity between products," admitted the president of an ad agency, "the less part reason really plays in brand selection." Huge amounts of money and talent tried to figure out what part reason and what part folly played in choosing one brand over another. Makers of every kind of product enlisted the research talents of advertising agencies, marketing companies, and sociologists to answer the question: what makes women buy? The effort to make women buy was so massive, a critic complained, that it was turning the poetry of life not to prose but to advertising copy.[43]

Television advertising was indispensable in the suburbs. The people living there consumed more, on average, than city dwellers. Suburbanites, after all, had more money and more kids, and they were furnishing new homes. They also required cars and other goods, like washing machines, that city people did not necessarily need. More importantly, most suburban dwellers no longer shopped at small and familiar stores on a pedestrian retail street. They made most of their purchases in the impersonal world of the shopping center. The title of a 1955 piece in *Fortune,* "Retailing: It's a New Ball Game," suggested the importance of the shift in shopping. Manufacturers began to pitch their brand-name goods directly to consumers through television and magazine ads, thereby assuming more of the selling services historically performed by small shopkeepers. Suburban retailers, in turn, had to lure their unknown customers with promotional flyers, newspaper inserts, local radio ads, and coupons and green stamps.[44]

This first generation of suburban women had to adjust to the surge in advertising and to the new kind of commerce it represented. That meant, above all, dealing with the loss of the retail street. We now take for granted what was big news in the '50s. "Instead of making daily rounds of Main Street," reported the *New York Times Magazine,* "a housewife now buys most of her needs in one place." Most magazine stories about new suburbs recorded a longing for the old street. "I wish there was some place close by to walk to, like the candy store in the city," complained one housewife. "Just some place to take the kids to buy a cone or newspaper in the afternoon. It helps break up the monotony of the day." Another woman wanted "to live where I can walk with my baby carriage and see stores instead of houses."[45]

The lack of a retail street accessible by foot also meant driving—everywhere. An iconic image of suburbs, and one which always impressed foreigners, was a woman behind the wheel of a wood-paneled station wagon loaded with kids and groceries. She drove along new commercial strips made to be reached by car. She drove by huge neon signs meant to fetch customers for McDonald's, Bob's Big Boy, and Howard Johnson's. And she drove by shopping centers: there were fifteen hundred of them in the United States in 1959; three years later, as suburbs kept growing, there were seven thousand.[46]

The shopping center was efficient, full of choices, and impersonal. It struck one writer as "a woman's dream world: an isolated, co-ordinated, well-designed, comfortable arrangement of nothing but shops and service." The typical shopper was a mother wearing slacks and a blouse, a simple dress, even shorts in summer. She shopped while pushing a stroller and keeping an eye on her other kids. The woman who shopped "in one of these impersonal marts" was said to be "as different from her grandmother or mother as the store itself is different from that of an earlier day."[47]

One difference was the lack of personal attention. Whereas the service butcher had once taught young housewives how to prepare various cuts of meat, women rarely saw butchers in the new supermarkets. The meat was already cut and conveniently packaged. Even managers of large stores noted the loss of personal service. Sears tried to rectify the loss by training its employees to be nice and friendly, just like the local storekeeper. Another difference was the endless variety of goods. Images we now take for granted were shocking in the '50s: the aisles in a supermarket looked like stacks in a giant food library; shoppers seemed vaguely robotic, thanks to Muzak; women at checkout counters were observed to be "lining up like cows at milking time—patient, not visibly expectant, the carts with food piled in rounded pyramids suggesting udders upside down."[48]

Community and Culture

Lawns, modern kitchens, and shopping centers fascinated observers of '50s suburbs. So did the intense nature of community life: "It is not altogether clear just why suburbanites want to be groupy," wrote John Keats in 1960. "Perhaps, being largely the generation now in its late thirties and early forties, they inherit a legacy of groupiness from the Depression days, or from the New Deal, or from the experience of the war, or, more prosaically, it may be because they share a feeling of inadequacy. Whatever the reason, there is no doubt that suburbia's watchword is 'Get with it.'"[49]

To say that suburbanites were especially groupy is saying a lot, because Americans are inveterate belongers. But community involvement in America may well have hit a high point in suburbs during the '50s. Since then, the time spent on informal socializing and visiting, or devoted to labor unions, churches, and clubs, or dedicated to political activities of all sorts, from attending a stump speech to sitting through a school-board meeting, has declined markedly.[50]

New suburbs certainly exhibited the frenzied participation that often enlivens new situations before they settle into familiar patterns. The early life of a suburb was so energetic, in fact, and social networks were assembled so quickly, that after just a few months residents felt involved and rooted. Most of them stayed involved, too, through all manner of religious, social, and community activities.[51] William Whyte exaggerated when he called the suburbs a "communal way of life," for there were plenty of bored, lonely, and isolated people. And the apathetic, as always, outnumbered the very active. But not by so much as elsewhere. The active, moreover, were so active that they generated a widespread feeling of activity. Some were even compelled to laugh at their own involvement. "Actually, neither Fred nor I are joiners, like some of these silly characters around here," explained one woman; "but it's gotten so now I practically have to make an appointment to see him Saturdays. During the week we alternate; when I have my meetings he baby-sits for me, and when he has his political meetings, I baby-sit for him."[52]

Most of the community's life revolved around children. There were so many of them, and they were so dictatorial, that one observer saw suburban life as a "filiarchy." The typical family had three or four children. The vast majority of kids (especially in the newest suburbs) were under fourteen years of age and many were under six.[53] Few of the younger ones went to day care, preschool classes, or kindergarten. Not many school-age kids spent summers at camps. Couples rarely employed outside help or had siblings or parents around to watch their kids on a regular basis. So a young mother spent most of her time caring for her children. She often watched them in the company of other mothers, and she may have taken part in what a magazine article on the new suburbs called a "kind of floating, day-long talk-fest, shifting from house to house."[54] There she talked with other mothers about pregnancy, how the kids were doing in school, what Dr. Spock had advised in *Baby and Child Care.* One woman remembers the community life of mothers as "a warm, boring, completely child-centered little culture. We sat around in each other's kitchens and backyards and drank a lot of coffee and smoked a million cigarettes and talked about our children. There was some competition, yes, but mostly we were young mothers and we were learning from each other." No wonder wives who had yet to have children were said to "suffer greatly from loneliness."[55]

In addition to watching their kids in kitchens and backyards, mothers (and a few fathers) also drove them to gatherings of Brownies, Cub Scouts, and dozens of other associations. Some parents hosted organizational meetings in their houses. Others chaperoned trips or donated their time to bake sales, cookie drives, and phone banks. From April to June, the lives of many boys and their parents revolved around Little League baseball. The number of leagues grew by nearly ten times during the '50s, when baseball was America's sport. Mothers drove their sons to games and watched them play. Fathers participated as fans, coaches, and umpires. Some of them raised money for uniforms and equipment from local supermarkets, manufacturers, oil-heating companies, and appliance, shoe, and furniture stores.[56]

While not as much fun as Little League, the PTA was the most famous, and perhaps the most important, of all suburban associations. The greatest school-building boom in the history of the nation probably occurred in suburbs during the '50s. Nearly all the money came from local property taxes and state coffers, so every PTA pressed its school board and state representatives for funding. It organized field trips for students and sponsored evenings for parents to meet teachers. It encouraged performances of public skits and plays by kids. At times it even got involved in curriculum decisions and the hiring (and firing) of school principals.[57]

Suburban organizations exemplified a trait that had interested visitor Alexis de Tocqueville more than a century earlier: the capacity of ordinary Americans to get things done on their own. In 1830, that meant repairing a bridge, establishing a church, or building a road. In the 1950s, it meant organizing a PTA, a Little League, the Jaycees, even a nursery school. In 1952, a mother called a meeting to address the need for a nursery. Fifty people came. "'My kitchen was just covered with rubbers and galoshes,' said dark-haired Betty. 'I never belonged to any organization before in my life and here I had helped start one. I got the feeling of people wanting to do things together.'" They found an old schoolhouse seven miles away. Husbands built chairs, toys, swings, blocks, easels, and "cubbies" for the children's clothes. The women published a town directory and sold advertising space in it to earn money.[58]

If most of the community's life revolved around kids, Sunday mornings were reserved for God. *Life* noticed "an unprecedented revival in religious belief and practice" after the war. Church membership increased from less than half of the population in 1940 to almost two-thirds by the middle of the '50s. The Bible topped the best-seller list in 1952, it was the leader in total book sales from 1952 to 1955, and it sold better in the '50s than in any other decade. The phrase "one nation under God" was added to the Pledge of Allegiance in 1954. "In God We Trust" was inscribed on paper money a year later.[59]

Religion was especially conspicuous in suburbs, which had most of the new churches and synagogues and the highest levels of community involvement in religious affairs. If the health of religion "be measured in numbers," wrote *Life*, "America's faith has never been more secure." Yet some complained of a "good fellowship and good works" kind of religion, of an approach to faith that was too practical, exceedingly casual, lacking in spiritual intensity, and merely in search of a sense of belonging. If that was true, it may have been because young couples hoped that membership in a church would offset the absence of extended family members, compensate for the high rate of turnover in some subdivisions, and make up for the friends, ethnic culture, and sense of community they had left behind in the old neighborhood or the small town. According to a city-bred person living in a new suburb, "You wanted to find people like yourself, and feel connected with people you knew something about, were sure of. The church was the place you turned to."[60]

Not everyone belonged to a new church, to one of the many groups centered on kids, or to a Jaycees that sponsored Fourth of July parades, soapbox derbies, and candidates for county government. But enough incentives existed to get a lot of people involved in community affairs of one sort or another. One incentive was to get something tangible in return for one's efforts. Nearly every couple owned a house and cared about property values. Almost every couple had children, or was about to have them, and so worried about the schools and the popularity of their kids. And just about every adult, as demonstrated in this conversation from McPartland's *No Down Payment*, had left behind family and friends and therefore needed new companions:

"None of us knew each other," Jean said. "Now we're neighbors, and we're more than neighbors. We're all the family most of us have out here."

"It is kind of strange," said Betty. "I was always used to knowing everybody in our block back home, and there were cousins, uncles, aunts—and of course my own family, brothers, a sister, Momma—"[61]

There were millions of others, wrote Frederick Lewis Allen in *Harper's*, who participated in community affairs simply because the concept of responsibility to the general public had become more and more widespread. The New Deal, the war effort, the union movement—Allen cited each as a reason for the growing sense of public obligation. One person even remembers a "terrible lust for premature maturity, this irresponsible desire for responsibility, before I had any idea what maturity involved or had ever tasted the pleasures of youthful irresponsibility." That sense of obligation helps explain why so many people joined local chapters of national organizations like the Rotary Club, the Red Cross, and the League of Women Voters, and why so many others volunteered at their local libraries and hospitals. Still other residents, according to Harry Henderson's 1953 study of suburbs, used their participation in local affairs to acquire status: "Since no one can acquire prestige through an imposing house, or inherited position, activity—the participation in community or group affairs—becomes the basis of prestige." Young adults assumed positions of leadership in suburbs, something they rarely did in a small town or city neighborhood. The most active among them became "big wheels" in the community.[62]

The high level of community activity made the average mother busier, and got her out of the house a great deal more, than her own mother: "What with cooking and keeping house and shopping and chauffeuring the children all over town," wrote an exasperated parent to her mother, "and trying to stay abreast of current affairs and going to meetings and helping out at the Red Cross, the hospital, the school cafeteria, etc., it seems as though I just don't have a second to myself any more."[63] It was an exaggeration to say, as one writer did, that "the self-reliant integrated woman is no longer exceptional and is accepted in the world of men."[64] But such statements did capture the sense of progress that so many women felt at the time. Mainstream magazines like *Coronet, Reader's Digest,* and *Ladies Home Journal* supported the idea that women, in addition to being good wives and mothers, should also strive to work, enter politics, volunteer in the community, and thereby compete in all fields. The proportion of magazine pieces that focused on motherhood, marriage, and housewifery was smaller in the '50s than it was in the '30s and '40s, while the number that praised successful women and encouraged women to engage in public life was greater. Mamie Eisenhower, during an award ceremony for successful women, said, "We can all take pride in the forward steps women have taken during our own generation to a role of leadership in community and even national affairs."[65]

Community meant more than groups and activities dedicated to God, children, and the social good. Some couples went to movies together. Others drove to downtown nightclubs or restaurants. Members of suburbia's upper crust enjoyed themselves at country clubs. But the single biggest recreational activity among adults was bowling. The number of lanes doubled to over a hundred thousand during the '50s. The new ones, built in the suburbs, were equipped with automatic pin-setters. In 1957, bowling took in more than ten times the amount of ticket sales for all major-league baseball games. Most people bowled as members of a team that competed in a league. Some of those teams were sponsored by unions or employers, others were supported by community groups and associations that used the alleys for fund-raisers, meetings, and membership drives. An alley owner who thought of installing a pool room decided against it because pool players, he said, were a "type of hoodlum," repellent to women bowlers especially.[66]

Young couples also spent a lot of time at each other's houses. They could not normally walk to bars, coffee shops, or restaurants. Their friends lived just a few houses down or a quick car ride away. Living rooms and kitchen-dining areas were perfect for entertaining. And many young couples liked to show off their houses. Visiting varied by class. White-collar and professional people, for example, probably entertained in their homes more often than working-class couples. And the martini was their drink of choice.

Their hair coloring of choice was Clairol's Nice 'n Easy. The ads, according to their creator, depicted a woman who was like "the proverbial girl on the block who's a little prettier than your wife and lives in a house slightly nicer than yours." In a famous "Does she or doesn't she?" commercial, a pretty suburban housewife with an apron over her cocktail dress prepares hors d'oeuvres: a dip made of sour cream and onion-soup mix, small square sandwiches made of white bread with the crust removed, and a cream cheese and olive spread on crackers. Her husband, in jacket and tie, comes into the kitchen for more gin and scotch. He gives her a kiss, pats her blonde hair, and watches proudly as she carries out the tray.[67] She takes the tray into the living room, where white-collar families liked to entertain.

Such a living room was open, sometimes sunken, or, if in a split level, a few steps below the kitchen and dining area. The room was said to be "clean looking" because it was open, free of molding, uncluttered by family heirlooms, and appointed with modern furniture in either the machine look or the very popular hand-crafted look called "Scandinavian modern." Modern furniture was light, set low to the ground, and supported by slender, slightly splayed, wooden legs. It was usually done in simple, single colors, and was occasionally lit by black pogo-stick lamps.

As a whole, it was probably America's best and most popular mass-produced furniture to date.[68]

Blue-collar couples, on the other hand, rarely entertained in the living room, which they saved for family pictures on Easter Sunday, a wedding day, or a graduation day. The living room also showed visitors that the house was properly furnished, usually with boxy, light-colored maple pieces in traditional Colonial designs, and occasionally with flat-slab sofas and coffee tables in the shape of amoebas or boomerangs.[69] The kitchen table was the center of activity, where two or three couples talked and played cards, drank beer or whiskey and water, and ate peanuts, chips, and pretzels. They might also have listened to records played on a living-room hi-fi system. After a few years of such gatherings, the smoke of a thousand cigarettes had stained brown the white swirled ceiling above the table.

Informal house visiting, the absence of older folks, the dominance of children, and a rather frenetic community life all helped to create a casual culture. Nothing better reflected the informality than clothes. Visitors from the city would say, "Why, nobody dresses around here!" Casual dress was fine for the shopping center, for visiting neighbors, for attending local meetings and children's social events. Casual meant fewer suits, hats, and dresses, and more sports jackets, knitted ties, jeans, shorts, capri pants, and sneakers. It also meant brighter colors. Striped and plaid sports jackets, canary-yellow shorts, slacks in pastel colors, honey-colored trousers—such finery inspired one critic to wonder whether this new style of dress was "part of the feminization of men, as some psychiatrists have called it, or whether this is just men reverting to a natural instinct to dress up in fancy plumage." The switch to casual dress took time and some getting used to. A neighbor saw the hostess of that night's cocktail party wearing bright capri pants. The neighbor mistook the pants for pajamas and assumed the party was called off.[70]

The "stiffness in collars," it was observed, "has gone the way of stiffness in manners."[71] Shopping in the downtown or along a neighborhood retail street still demanded a measure of decorum. In suburbs, on the other hand, the young mother got into her car, usually with her kids, and drove to shopping centers where she saw other young mothers with children. Whereas neighborhood shopkeepers, along with nearby grandparents, aunts, and uncles, imposed a certain control on children, suburban kids were indulged by their parents and most neighbors. The house or apartment in the city served as a sanctuary from the street. The suburban yard, by contrast, was practically an extension of the house, neighbors often visited each other without warning, and every person's comings and goings were plain to see.

"Gone also," concluded an observer, "are most rituals and ceremonies" that concern how one gets to know other people, establish friendships, and form groups. The newness of it all, the great number of things to get organized, the lack of elders to impose a certain restraint and formality—all fostered a spontaneity and casualness rarely known in the city. Residents were polite, of course, but they showed less of the distance, caution, and discretion associated with urban life.[72]

Relaxed manners and a high rate of participation in community groups did not mean everyone got along. Because so much revolved around the children, almost every woman found herself associating with at least a few other women—PTA members, volunteers in children's groups, mothers of her own kids' friends—whom she did not like. Women formed cliques as well. A woman in a Chicago suburb reported the following episode, which must have had multiple renditions throughout the country. A certain Estelle

was dying to get in with the gang when she moved in. She is a very warmhearted gal and is always trying to help people, but she's, well—sort of elaborate about it. One day she decided to win over everybody by giving an afternoon party for the gals. Poor thing, she did it all wrong. The girls turned up in their bathing suits and slacks, as usual, and here she had little doilies and silver and everything spread around. Ever since then it's been almost a planned campaign to keep her out of things. Even her two-year-old daughter gets kept out of the kids' parties.[73]

To be unpopular is a great tragedy in America, and it was especially so in suburbs. The extended family was absent, and ostracized couples had to see neighbors getting in and out of their cars, hanging out clothes, mowing lawns, and entertaining visitors.

Resentment and pettiness were as common in suburbs as anywhere else. One man bought a dog, he said, "because I am damn sick and tired of my neighbor's dog yapping all night. I just want to give them a taste of what it's like." Many couples learned to dislike each other because their kids did not get along. Others had to tolerate each other because their kids did get along. Some couples became friends immediately only to realize, a little later, that they had nothing in common.[74]

The culture of suburbs may not have guaranteed harmony, but its groupiness, its basis in home ownership, its economic and social equality, and its focus on kids and consumer items did put a "premium," it was said in *Harper's*, "on a kind of amiable, thoughtless conformity." Such conformity worried critics who felt that everyone "would become just like everyone else—a thin slice of yellow plastic cheese in the long, soft loaf of Velveeta that was America." Adlai Stevenson warned of mass mediocrity.[75]

Some critics even saw residents of suburbs as dupes. To writer John Keats, for example, suburbanites say to themselves, " 'We never had it so good.' Then off they'll go, drifting from one vague disappointment to the next, deep in the narcotic trance of advertised promise, never thinking of themselves but always of their diversions, entirely unaware that they are neither giving nor receiving anything of value."[76] Most of this criticism was misplaced carping by snobs who were resentful of losing their status as arbiters of good taste to an expanding mass market. Much of it was based on a superficial understanding of suburban life. Some of it was on the mark.

Suburban life, for example, did stress conformity in an age already distinguished by its coherence. Having three or four kids was as much a duty as a choice. The pressure to join groups was stronger in suburbs than anywhere else and possibly stronger in the 1950s than at any other time. Residents stayed within strict bounds regarding the decoration of their houses, the care of their yards, and the kind of car, clothing, and furniture they bought. They not only kept up with the Joneses, observed the *New York Times,* but down with them as well.[77]

But why would anyone *not* follow the rules, accept the codes, and buy the entire package of suburban life? The typical couple, after all, owned their own house and thus cared deeply about the appearance and value of their property. They saw their incomes grow almost every year, took summer vacations, and watched their healthy kids leave each morning for a safe and good school. They trusted a government that guaranteed their mortgages, put veterans through college, built sewer systems and highways, and vaccinated their kids at the local school or fire station. These were young parents who had grown up during the Depression and so wanted better lives for their kids. Many of the fathers had served in a war that generated national pride. Everyone understood that a cold war was being fought against an enemy whose way of life (and nuclear arms) was a threat to their own. And a lot of residents relied on the community life of their suburbs for a sense of belonging.

In a large survey of married couples, over 80 percent of husbands and wives rated their marriages "above average," and nearly two out of three saw their own as "extraordinarily happy" or "decidedly happier than average."[78] A few women may have responded that way to feel good about themselves. Some may have been putting up a brave front. Others, as

Betty Friedan wrote in *The Feminine Mystique,* may have been unable to make sense of an embryonic feeling of unease:

> It was a strange stirring, a sense of dissatisfaction, a yearning that women suffered in the middle of the twentieth century in the United States. Each suburban wife struggled with it alone. As she made the beds, shopped for groceries, matched slipcover material, ate peanut butter sandwiches with her children, chauffeured Cub Scouts and Brownies, lay beside her husband at night—she was afraid to ask even of herself the silent question—"Is this all?"[79]

Yet the "all" that left Friedan so unsatisfied did satisfy most other young mothers. As a whole, they were better off economically than any previous generation. They participated in public life more than their mothers or grandmothers did. Their husbands involved themselves in family life like their own fathers never had. Parents, neighbors, and experts assured them that they were doing the right thing. The popular press, all the while, covered their situation in great detail: "Caught as she is in conflicting currents," reported *Life* in 1956, "the American woman of today finds herself being analyzed and admired, envied and criticized as never before."[80]

It is no surprise, however, that many more women than men were ambivalent about their marriages and their lives. Some suburban women felt disconnected—from a career, from a local retail street, from extended family members. Most of those who did feel uneasy or isolated compensated by focusing on their kids or getting involved in community activities. Some had affairs. A few drank. But the prevailing belief at the time was to make the family work. After recalling a dull life in her suburb, a middle-class woman who did not work outside of the house said, "But I have to be fair, here. As dissatisfied as I was, and as restless, I remember so well this feeling we had at the time that the world was going to be your oyster. You were going to make money, your kids were going to go to good schools, everything was possible if you just did what you were supposed to do. The future was rosy. There was a tremendous feeling of optimism."[81]

The level of conformity may have been high, but it was not unwilling. Nor was it blind to its own faults. Most people who saw the suburbs as their chance for a good life also knew that the good life was no paradise. Except for magazine ads and situation comedies, the mass media told them as much. *Fortune,* for example, wrote about "the new masses," while the *Saturday Evening Post* reported on "trouble in the suburbs."[82] Widely read books like *The Organization Man, The Lonely Crowd,* and the *Hidden*

Persuaders warned specifically of conformity. Popular novels like *The Man in the Gray Flannel Suit* and *The Crack in the Picture Window* revealed dark sides of suburban life. Even Hollywood, on occasion, gave the suburbs an edge. In *Pitfall,* for example, Dick Powell says he feels like "a wheel within a wheel within a wheel." He has a pretty wife, a little boy, a nice house, and a good job downtown. But he strays into an affair, is blackmailed, and nearly loses his family.

All of this apparent conformity went along with what some decried as "mass mediocrity." Writers like William Whyte (in *The Organization Man*) and C. Wright Mills (in *White Collar* and *The Power Elite*) blamed the mediocrity on the dulling effect that large companies, government bureaucracies, and even labor unions were thought to have on workers and citizens. Most others blamed the mass culture that dominated the new suburbs. "Qualitatively," wrote *Life* in 1954, "it may or may not yet be great culture, but quantitatively there is an awful lot of it; with culture as with everything else they undertake, Americans have to have more of it than anybody else." Five years later, after praising America's "smashing success politically and economically," the magazine still was uneasy about "mass culture. Can such a sprawling, classless society," it asked, "support the high standards of quality and achievement that aristocratic ages bequeathed us?"[83]

That was a ridiculous question. For the culture of suburban life—the food, the clothes, the shopping centers, the dominance of children, the paintings hung on living-room walls, the time spent watching television—was not refined. But the growing middle-class culture in suburbs did not displace so-called higher standards of culture. There was, in fact, more of every kind of culture for anyone who wanted some—from symphony orchestras and art museums to rock-and-roll records and romance novels.[84]

The most extreme complaint about suburbs concerned the gap that John Keats saw between "the narcotic trance of advertised promise" and the banal realities of everyday suburban life. People were deluded, in other words, and lived empty lives. Even the *Saturday Evening Post* acknowledged a grain of truth in the complaint that "the spread of suburbia is stultifying the American mind." Yet the magazine also understood that "the man who spends a pleasant Saturday puttering about his petunias, or mowing the lawn, or building a boat in his garage, or sitting with a beer before the television set, watching a baseball game, is living as he likes to live."[85] Furthermore, there is no evidence that suburbanites enjoyed themselves any less, participated any less in civic life, or were one bit less responsible as parents, workers, or citizens, than people who lived in the old neighborhoods or in small towns.[86] If the suburban home-owner was beguiled by an asinine television commercial for a '55 Chevy, what about today's thirty-year-old who lives in a gentrified neighborhood and thinks the latest ad for a cell phone or a sports utility vehicle is, like, cool?

The Suburbanization of Everything

The average suburb relied on its city well into the 1950s. In 1958, for example, one writer depicted suburbs as "culturally and economically dependent upon the central city," while another characterized them as "oriented toward" the cities.[87] They saw it that way because, until the middle of the '50s, city factories made most of the products used to build suburbs and supply their residents, many suburban dwellers worked in cities and occasionally shopped and enjoyed themselves downtown, and a lot of suburban dwellers grew up in cities and visited parents and siblings in the old neighborhoods.[88]

By around 1960, however, observers agreed that suburbs were becoming dominant economic areas in their own right. More than half of the people living in the twenty largest metropolitan areas now resided in suburbs. Fewer suburbanites worked in cities. Suburban factories and warehouses took business from central city manufacturers and shippers. Businessmen took cars and airplanes instead of trains. Travelers stayed in suburban motels rather than city hotels. Shopping centers, meanwhile, had undermined downtown's role as the retail center of the metropolis.[89]

Suburbs disengaged from the culture of cities, too. Although many couples went downtown in the '50s for its bars, theaters, and nightclubs, residents increasingly stayed in the suburbs to enjoy themselves. Young parents who grew up in cities quickly adopted a casual way of life that revolved around kids, cars, yards, shopping centers, and community groups and associations. And most suburbs became separate political entities that successfully resisted annexation by their central cities.

Karl Marx described nineteenth-century capitalism as "the urbanization of the countryside," because he saw cities spreading their machines, markets, and money across the land.[90] A main theme of American capitalism since the 1950s, he might say today, has been "the suburbanization of everything." For if suburbs at first depended upon and then became detached from and even antithetical to their cities, they eventually sent their distinctive features back into those cities: dogs now run rampant in city parks; McDonald's sells burgers downtown; everyone dresses casually; neighborhood people drive to supermarkets; new public-housing projects feature duplexes with tiny yards; old clusters of corner stores give way to little shopping centers built around parking lots; developers hang balconies from the sides of converted warehouses so residents can grill meat; new high-rise apartment buildings feature underground parking garages and built-in health clubs, minimarts, and drugstores so their residents don't have to use the streets. If America reached its peak as an urban civilization in the 1950s, it has since become a quintessentially suburban one—even in its cities.

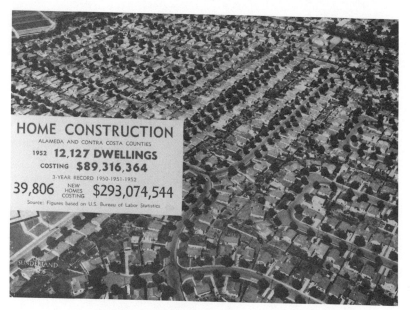

HOME CONSTRUCTION
ALAMEDA AND CONTRA COSTA COUNTIES
1952 **12,127 DWELLINGS**
COSTING **$89,316,364**
3-YEAR RECORD 1950-1951-1952
39,806 NEW HOMES COSTING $293,074,544
Source: Figures based on U.S. Bureau of Labor Statistics

FIGURE 42. Aerial shot of new subdivision from a monthly magazine called *Bright Spots,* published by the *Oakland Herald Tribune* to document and encourage the growth of housing, retail sales, and new factories in the greater metropolitan area. Courtesy of Oakland Public Library, History Room.

FIGURE 43. The new Stonestown Shopping Center in suburban San Francisco, 1953. Courtesy of San Francisco History Center, San Francisco Public Library.

FIGURE 44. A couple with their first child in a new Tucson suburb, 1951. Courtesy of Michael Solot.

FIGURE 45. The same Tucson couple in front of their house with their children, ready to drive to the airport for the start of a family vacation, early 1960s. Courtesy of Michael Solot.

FIGURE 46. Hoover vacuum cleaner ad from *Life*, April 21, 1958 (p. 63), an example of what *Fortune* called the "upheaval in home goods." Women in ads for domestic goods typically wore dresses and high heels. Courtesy of the Hoover Company.

MAN WITH A PLAN. By following the simple steps of a Scotts Program, he'll have a lawn the whole neighborhood will envy.

FIGURE 47. Ad showing suburban man tending his lawn. Courtesy of The Scotts Company, Marysville, Ohio, owner of the trademarks "Scotts" and "Turf Builder."

Notes

Chapter 1. The City at Its Peak

1. Jan Morris, *Manhattan '45* (New York: Oxford University Press, 1987), 12; James Vance, conversation with author (Berkeley, Calif., September 1997); see also Raymond Murphy and James Vance, "Delimiting the CBD," *Economic Geography* (July 1954); Raymond Murphy and James Vance, "A Comparative Study of Nine Central Business Districts," *Economic Geography* (October 1954).

2. Harold Martin, "Is Downtown Doomed?" *Saturday Evening Post* (January 9, 1960); see also Raymond Vernon, *The Changing Economic Function of the Central City* (New York: New York Metropolitan Region Study, 1959), 44; Edgar Hoover and Raymond Vernon, *Anatomy of a Metropolis* (Cambridge: Harvard University Press, 1959), 116–21; George Sternlieb, *The Future of the Downtown Department Store* (Cambridge: MIT Press, 1962), 67, 84, 113, 157.

3. Joseph Clark, "A Voice for the Cities," *Nation* (March 7, 1959); Martin, "Is Downtown Doomed?"; see also editors of *Fortune, The Exploding Metropolis* (Garden City, N.Y.: Doubleday, 1958), 53; "What's Happening to U.S. Cities?" *U.S. News and World Report* (June 20, 1960); Christopher Tunnard, "America's Super-Cities," *Harper's* (August 1958); Harold Martin, "Are We Building a City 600 Miles Long?" *Saturday Evening Post* (January 2, 1960); Sternlieb, *Future of the Downtown*, 15, 84; Nathan Glazer, "Megalopolis and How It Grew," *Reporter* (November 12, 1959); William O'Hallaren, "A Fair Share for the Cities," *Reporter* (November 12, 1959).

4. Cited in Stephen Farley, Regina Kelly, and the Ward 6 Youth History Team, eds., *Snapped on the Street: A Community of Photos and Memories from Downtown Tucson, 1937–1963* (Tucson, Ariz.: Tucson Voices Press, 1999), 57.

5. E. B. White, *Here Is New York* (1949; reprint, New York: Harper and Row, 1977), 24–25.

6. Cited in Jervis Anderson, *This Was Harlem* (New York: Farrar, Straus and Giroux, 1982), 346.

7. Howell Walker, "Cities Like Worcester Make America," *National Geographic* (February 1955), 189–214; William B. Dickinson, *This Is Greater Philadelphia* (Philadelphia: Bulletin Company, 1954), 3; see also E. Willard Miller, *A Geography of Manufacturing* (Englewood Cliffs, N.J.: Prentice-Hall, 1962), 31; Vernon, *Changing Economic Function,* 49–51, 70–73, 74–79; Harold Mayer and Richard Wade, *Chicago: Growth of a Metropolis* (Chicago: University of Chicago Press, 1969), 426, 428; Walter Guzzardi, "Freight Goes Forward with Forgash," *Fortune* (July 1962); Paul Jacobs, *The State of the Unions* (New York: Atheneum, 1963), 91.

8. Pete Hamill, *A Drinking Life: A Memoir* (Boston: Little, Brown, 1994), 60; Geraldine Fregoso, "Growing Up in the Mission," *San Francisco Sunday Examiner and Chronicle* (April 21, 1974); Marya Mannes, *The New York I Know* (New York: J. B. Lippincott, 1961), 64–65.

9. Martin, "Are We Building a City 600 Miles Long?"

10. Philip Roth, interview by Terry Gross, *Fresh Air,* National Public Radio (May 8, 2000).

Chapter 2. The Downtown

1. Raymond Murphy and James Vance, "A Comparative Study of Nine Central Business Districts," *Economic Geography* (October 1954), 301.

2. James Vance, *The Continuing City* (Baltimore: Johns Hopkins University Press, 1990), 465.

3. Earle Schultz and Walter Simmons, *Offices in the Sky* (New York: Bobbs-Merrill, 1959), 229–31; Edgar Horwood and Ronald Boyce, *Studies of the Central Business District and Urban Freeway Development* (Seattle: University of Washington Press, 1959), 69–70, 93; Edwin Goldfield, *Statistical Abstract of the United States, 1965* (Washington, D.C.: U.S. Government Printing Office, 1965), 227.

4. Russell Lynes, *A Surfeit of Honey* (New York: Harper and Brothers, 1957), 72–73.

5. Norman Podhoretz, cited in Barbara Ehrenreich, *The Hearts of Men: American Dreams and the Flight from Commitment* (Garden City, N.Y.: Anchor, 1983), 41; Ben Fairless, cited in *Life* (October 22, 1956). On corporate growth and culture, see William Whyte, "The Wives of Management," *Fortune* (October 1951); John Brooks, *Seven Fat Years: Chronicles of Wall Street* (New York: Harper, 1958); John Brooks, *The Great Leap* (New York: Harper, 1966), 38–72; William Whyte, *The Organization Man* (New York: Anchor, 1957).

6. Whyte, *The Organization Man,* 73.

7. Clyde Kluckhohn, "Has There Been a Discernible Shift in American Values During the Past Generation?" in Elting Morison, ed., *The American Style* (New York: Harper and Row, 1958), 155–56; Charles Silberman and Sanford Parker, "How the U.S. Can Get 50 per Cent Richer," *Fortune* (March 1959); Whyte, "Wives of Management"; William Whyte, "The Corporation and the Wife," *Fortune* (November 1951).

8. Ben Wattenberg, *This U.S.A.: An Unexpected Family Portrait of 194,067,296 Americans Drawn from the Census* (Garden City, N.Y.: Doubleday, 1965), 168;

quote from Arturo Gonzalez and Janeann Gonzalez, "Where No Woman Reaches the Summit," *New York Times Magazine* (August 17, 1958), 34.

9. John Marquand, *Point of No Return* (Boston: Little, Brown, 1949), 30; George Gallup, *The Gallup Poll: Public Opinion, 1935–1971* (New York: Random House, 1972), 2: 1061–62; see also Shepherd Mead, "Analysis of the Genus Secretarius," *New York Times Magazine* (April 22, 1956).

10. Katherine Hamill, "Women as Bosses," *Fortune* (June 1956); Gonzalez and Gonzalez, "Where No Woman."

11. Jerome Weidman, *The Enemy Camp* (New York: Random House, 1958), 3.

12. Margot Patterson Doss, *San Francisco at Your Feet* (New York: Grove Press, 1964), 33.

13. Max Hess, *Every Dollar Counts: The Story of the American Department Store* (New York: Fairchild, 1952), 164; Dero Saunders, "Department Stores: Race for the Suburbs," *Fortune* (December 1951), 168.

14. Quotes from Saunders, "Department Stores," 168, 99; Gerald Weales, "Small-Town Detroit," *Commentary* (September 1956). On retail sales, see Horwood and Boyce, *Central Business District*, 44; William Dobriner, ed., *The Suburban Community* (New York: G. P. Putnam, 1958), 199; on suburbanites shopping downtown, see "The Lush New Suburban Market," *Fortune* (November 1953); Hal Burton, "Downtown Isn't Doomed!" *Saturday Evening Post* (June 5, 1954); Schultz and Simmons, *Offices in the Sky*, 228; C. T. Jonassen, *The Shopping Center versus Downtown* (Columbus: Ohio State University, 1955); Economic Research Council, *Business Stability and Opportunities for Growth in the Syracuse Area* (Syracuse, N.Y.: Economic Research Council, 1954), 13; on problems of downtown department stores, see Charles Silberman, "The Department Stores Are Waking Up," *Fortune* (July 1962), 144; Raymond Vernon, *The Changing Economic Function of the Central City* (New York: New York Metropolitan Region Study, 1959), 44; George Sternlieb, *The Future of the Downtown Department Store* (Cambridge: MIT Press, 1962), 35, 113, 157; Adam Cohen and Elizabeth Taylor, *American Pharaoh* (Boston: Little, Brown, 2000), 164–65; Jon Teaford, *The Twentieth-Century American City* (Baltimore: Johns Hopkins University Press, 1986), 111; Robert Stern, Thomas Mellins, and David Fishman, *New York, 1960* (New York: Monacelli Press, 1995), 908–10.

15. William B. Dickinson, *This Is Greater Philadelphia* (Philadelphia: Bulletin Company, 1954), 25; see also Sternlieb, *Future of the Downtown*, 67.

16. Susan Porter Benson, *Counter Cultures: Saleswomen, Managers, and Customers in American Department Stores, 1890–1940* (Chicago: University of Illinois Press, 1986), 1–2.

17. Sternlieb, *Future of the Downtown*, 28; Jonassen, *Shopping Center versus Downtown*, 91.

18. Vance, *The Continuing City*, 467; Murphy and Vance, "Comparative Study"; F. W. Woolworth Company, *Woolworth's First 75 Years, 1879–1954* (New York: William Rudge's Sons, 1954), 31.

19. Ada Louise Huxtable, "The Death of the Five-and-Ten," in *Architecture, Anyone?* (Berkeley: University of California Press, 1986), 315.

20. Woolworth, *Woolworth's First 75 Years*.

21. Quote from Murphy and Vance, "Comparative Study," 335; see also Jonassen, *Shopping Center versus Downtown*, 27-43; Horwood and Boyce, *Central Business District*, 120.

22. Sternlieb, *Future of the Downtown*, 15; Economic Research Council, *Growth in the Syracuse Area*, 14; Jonassen, *Shopping Center versus Downtown*, 38-42, 95; "Shopping Centers," *Architectural Record* (October 1953), 178-201; Vance, *The Continuing City*, 465-67; Schultz and Simmons, *Offices in the Sky*, 229.

23. Goldfield, *Statistical Abstract*, 578; editors of Fortune, *The Exploding Metropolis* (Garden City, N.Y.: Doubleday, 1958), 54, 58; Edward Banfield, *Political Influence* (New York: Free Press, 1961), 91-92; Vernon, *Changing Economic Function*, 46; Sternlieb, *Future of the Downtown*, 60; Jonassen, *Shopping Center versus Downtown*, 31-33; "What's Happening to U.S. Cities?" *U.S. News and World Report* (June 20, 1960); Stanley Berge, "How Commuters Can Have Their Trains," *Atlantic Monthly* (May 1960).

24. Saundra Sharp, "Growing Up Integrated," in Elena Featherston, ed., *Skin Deep* (Freedom, Calif.: Crossing Press, 1994), 31.

25. St. Clair Drake and Horace Cayton, *Black Metropolis* (1945; reprint, Chicago: University of Chicago Press, 1993), 825.

26. Cited in David Halberstam, " 'A Good City Gone Ugly,' " *Reporter* (March 31, 1960), 18.

27. Miles Colean, *Renewing Our Cities* (New York: Twentieth Century Fund, 1953), 42; see also Raymond Murphy, James Vance, and Bart Epstein, "Internal Structure of the CBD," *Economic Geography* (January 1955), 41-45.

28. Harold Vatter, *The U.S. Economy in the 1950s* (New York: W. W. Norton, 1963), 19, 266; E. Willard Miller, *A Geography of Manufacturing* (Englewood Cliffs, N.J.: Prentice-Hall, 1962); Joshua Freeman, *Working Class New York* (New York: New Press, 2000), 3-200.

29. Goldfield, *Statistical Abstract*, 559; "Something Haywire in the Freight Business," *Fortune* (January 1957), 104; John Hess, "The Railroads Punish the Passenger," *Atlantic Monthly* (July 1957); Edward Thompson, "What Hope for the Railroads?" *Fortune* (February 1958); Miller, *Geography of Manufacturing*, 51-52; Edgar Hoover and Raymond Vernon, *Anatomy of a Metropolis* (Cambridge: Harvard University Press, 1959), 37; Murphy, Vance, and Epstein, "Internal Structure," 43.

30. Richard Gutman and Elliot Kaufman, *American Diner* (New York: Harper and Row, 1979), 44-54.

31. Leonard Wallock, *New York: Culture Capital of the World, 1940-1965* (New York: Rizzoli, 1988), 32; Benjamin Chinitz, *Freight and Metropolis* (Cambridge: Harvard University Press, 1960), 7; Jan Morris, *Manhattan '45* (New York: Oxford University Press, 1987), 268.

32. Doss, *San Francisco*, 43.

33. Female observer's quote from ibid.; second quote from Paul Jacobs, *The State of the Unions* (New York: Atheneum, 1963), 98.

34. Jacobs, *State of the Unions*, 22.

35. Dickinson, *This Is Greater Philadelphia*, 8.

36. The word "sinister" is from Doss, *San Francisco*, 43; the last quote is from

A. H. Raskin, "C-Men on the Waterfront," *New York Times Magazine* (October 9, 1955), 15. On dock corruption, see also Bernard Nossiter, "The Teamsters: Corrupt Policemen of an Unruly Industry," *Harper's* (May 1959); Clark Mellenhoff, "The Teamsters Defy the Government," *Atlantic Monthly* (November 1958); Mary Heaton Vorse, "The Pirate's Nest of New York," *Harper's* (April 1952); Sanford Gottlieb, "The Man Who Shut Down the Port of New York," *Reporter* (April 27, 1954); A. H. Raskin, "How the Docks Shape Up Now," *New York Times Magazine* (June 12, 1955); Robert Whalen, "Two Generals Patrol the Docks," *New York Times Magazine* (November 29, 1953); Budd Schulberg, "How One Pier Got Rid of the Mob," *New York Times Magazine* (September 27, 1953); A. H. Raskin, "Union Leader—and Big Business Man," *New York Times Magazine* (November 15, 1953).

37. Department of City Planning, *Central Commercial and Peripheral Area (Report No. 5)* (Chicago: Department of City Planning, 1954), 9–10; Vernon, *Changing Economic Function,* 35; Peter Muller, *Contemporary Suburban America* (Englewood Cliffs, N.J.: Prentice-Hall, 1981), 135, 148; "Is Downtown Doomed?" *Saturday Evening Post* (January 1, 1960).

38. Fred Frailey, *Twilight of the Great Trains* (Waukesha, Wisc.: Kalmbach, 1998), 5; Goldfield, *Statistical Abstract,* 583, 580; J. Hess, "Railroads Punish"; Gallup, *Gallup Poll,* 2: 1287–88.

39. Frailey, *Twilight,* 5; J. Ronald Oakley, *God's Country: America in the Fifties* (New York: W. W. Norton, 1986), 20.

40. Richard Cook, *The Twentieth Century Limited, 1938–1967* (Lynchburg, Va.: TLC Publishing, 1993); Morris, *Manhattan '45,* 174.

41. Will Stevens, *Three Street* (Garden City, N.Y.: Doubleday, 1962); John Bartlow Martin, "How Corrupt Is Chicago?" *Saturday Evening Post* (March 31, 1951); Weales, "Small-Town Detroit," 219–26; Horwood and Boyce, *Central Business District,* 23; Donald Bogue, *Skid Row in American Cities* (Chicago: University of Chicago Press, 1963), 1–15; Meyer Berger, "The Bowery Blinks in the Sunlight," *New York Times Magazine* (May 20, 1956); Raymond Murphy and James Vance, "Delimiting the CBD," *Economic Geography* (July 1954), 189–222; Murphy, Vance, and Epstein, "Internal Structure," 21–46; Vance, *The Continuing City,* 466–67; Sara Harris, *Skid Row, U.S.A.* (Garden City, N.Y.: Doubleday, 1956).

42. Jack Lait and Lee Mortimer, *New York: Confidential* (New York: Dell, 1951), 13–14; William Klein, *New York, 1954–1955* (Manchester, Eng.: Art Publishers, 1995), 7.

43. Eve Merriam, *Figleaf* (New York: J. B. Lippincott, 1960), 61.

44. Caroline Rennolds Milbank, *New York Fashion: The Evolution of American Style* (New York: Harry Abrams, 1989), 175; Merriam, *Figleaf,* 14; quote from Sygne, "A Frenchwoman in New York," *Reporter* (August 1, 1950), 37; see also Pierre Boulat, *Life* (September 16, 1957).

45. Merriam, *Figleaf,* 14; see also *Life* (August 31, 1953; September 3, 1956; and March 4, 1957).

46. R. Turner Wilcox, *The Mode in Costume* (New York: Scribner's, 1958), 293; *Cosmopolitan* (November 1955); see also Milbank, *New York Fashion,* 131–43; Merriam, *Figleaf,* 25.

47. Merriam, *Figleaf,* 14; Milbank, *New York Fashion,* 175.

48. Anne Hollander, *Sex and Suits* (New York: Alfred A. Knopf, 1994), 61.

49. Quoted in Brett Harvey, *The Fifties: A Women's Oral History* (New York: HarperCollins, 1993), xi.

50. Merriam, *Figleaf,* 209.

51. These ads were in *Life* and *Cosmopolitan* throughout the '50s; see also Merriam, *Figleaf,* 149.

52. *Life* (September 14, 1953; see also April 18, 1955; August 11, 1952; and March 24, 1958).

53. Hollander, *Sex and Suits,* 147–48.

54. Christopher Pearce, *Fifties Source Book* (London: Quarto, 1990), 145, 154–55; Milbank, *New York Fashion,* 176, 177, 179; quote from Merriam, *Figleaf,* 198–99, 98.

55. Thomas Hine, *Populuxe* (New York: Alfred A. Knopf, 1986), 86–87; see also Edward Weeks, "How Big Is One," *Atlantic Monthly* (August 1958); Alfred Sinks, "Those Big Fat Cars," *Harper's* (April 1949); Raymond Loewy, "Jukebox on Wheels," *Atlantic Monthly* (April 1955); *Life* (January 30, 1956).

56. The ads are from the *Saturday Evening Post* (January 20, 1951; February 10, 1951; March 3, 1952; and October 25, 1952); *Fortune* (November 1952 and April 1953); and *Life* (January 9, 1950; and September 26, 1955). On Dagmars, see Hine, *Populuxe,* 93–94; "Dagmar Is Her Name," *Cosmopolitan* (February 1951); "A New Kind of Car Market," *Fortune* (September 1953).

57. Hubert de Givenchy, cited in Merriam, *Figleaf,* 47–48; see also Wilcox, *The Mode in Costume,* 417; *Life* (September 16, 1957; and January 20, 1958); Gallup, *Gallup Poll,* 2: 1551; quote from man cited in John Fenton, *In Your Opinion* (Boston: Little, Brown, 1960), 188.

58. Mary Cantwell, *Manhattan, When I Was Young* (New York: Penguin, 1996), 20.

59. Milbank, *New York Fashion,* 172.

60. Robert O. Blood and Donald M. Wolfe, *Husbands and Wives: The Dynamics of Married Living* (New York: Free Press, 1960), 172; Paul Glick, *American Families* (New York: Wiley and Sons, 1957), 104, 54; Joseph Veroff, Elizabeth Douvan, and Richard Kulka, *The Inner American: A Self-Portrait from 1957–1976* (New York: Basic Books, 1981), 140–93.

61. Benita Eisler, *Private Lives: Men and Women of the Fifties* (New York: Franklin Watts, 1986), 110; Alfred Hitchcock quoted by Louis Jordan in *Grace Kelly: The American Princess,* Public Broadcasting Service special.

62. Alfred Kinsey, *Sexual Behavior in the Human Female* (Philadelphia: W. B. Saunders, 1953), 331; Elaine Tyler May, *Homeward Bound: American Families in the Cold War Era* (New York: Basic Books, 1988), 120–21.

63. Laurence Olivier, cited in David Halberstam, *The Fifties* (New York: Ballantine, 1993), 565; last quote from Harvey, *The Fifties,* xi.

64. Hollander, *Sex and Suits,* 166–67.

65. Yip Harburg, cited in Deena Rosenberg, *Fascinating Rhythm: The Collaboration of George and Ira Gershwin* (New York: E. P. Dutton, 1991), xvii.

66. Ira Gershwin, cited in Philip Furia, *The Poets of Tin Pan Alley: A History of America's Great Lyricists* (New York: Oxford University Press, 1990), 12.

67. Irving Berlin, cited in Philip Furia, *Irving Berlin: A Life in Song* (New York: Schirmer, 1998), 67. On the popular song, see Charles Hamm, *Yesterdays: Popular Song in America* (New York: W. W. Norton, 1979), 326–90; Alec Wilder, *American Popular Song: The Great Innovators, 1900–1950* (New York: Oxford University Press, 1972); Furia, *Poets of Tin Pan Alley;* Rosenberg, *Fascinating Rhythm.*

68. Irving Berlin, cited in John Lahr, "Revolutionary Rag," *New Yorker* (March 8, 1999), 78; George Gershwin, cited in Rosenberg, *Fascinating Rhythm,* 126–27; Hamm, *Yesterdays,* 378.

69. Berlin, cited in Furia, *Irving Berlin,* 43, 4.

70. Ibid., 43; H. L. Mencken, *The American Language,* 4th ed. (New York: Alfred A. Knopf, 1937), 91; Rosenberg, *Fascinating Rhythm,* xx–xxi, 31, 160–61.

71. George Gershwin, cited in Rosenberg, *Fascinating Rhythm,* xxvii.

72. Tony Bennett, in the liner notes to his *Steppin' Out,* Columbia compact disc CK57424 (1993).

73. Peter Bogdanovich, cited in John Rockwell, *Sinatra: An American Classic* (New York: Random House, 1984), 22; Davis, cited in Bill Zehme, "And Then There Was One," *Esquire* (March 1996).

74. *Newsweek* (May 25, 1998), 70.

75. Ibid., 71.

76. On sincerity, see Nora Sayre, *Previous Convictions: A Journey Through the 1950s* (New Brunswick, N.J.: Rutgers University Press, 1995), 138; David Riesman with Nathan Glazer and Reuel Denney, *The Lonely Crowd: A Study of the Changing American Character* (Garden City, N.Y.: Doubleday, 1953), 224; Frank Sinatra, cited in Henry Pleasants, *The Great American Popular Singers* (New York: Simon and Schuster, 1974), 196.

77. Thomas Pryor, "Rise, Fall, and Rise of Sinatra," *New York Times Magazine* (February 10, 1957), 17.

78. Yip Harburg, cited in Philip Furia, *Ira Gershwin: The Art of the Lyricist* (New York: Oxford University Press, 1996), 15; Furia, *Poets of Tin Pan Alley,* 97, 153–80.

79. Harold Mayer and Richard Wade, *Growth of a Metropolis* (Chicago: University of Chicago Press, 1969), 450–60; Allan Temko, "San Francisco Rebuilds Again," *Harper's* (April 1960); Sally Woodbridge and John Woodbridge, *San Francisco Architecture* (San Francisco: Chronicle, 1992), 31, 33, 35; "Pittsburgh Rebuilds," *Fortune* (June 1952); Karl Schriftgiesser, "The Pittsburgh Story," *Atlantic Monthly* (May 1951); Stefan Lorant, *Pittsburgh: The Story of an American City* (Garden City, N.Y.: Doubleday, 1964), 450–51; Carl Abbott, *The New Urban America: Growth and Politics in the Sunbelt Cities* (Chapel Hill: University of North Carolina Press, 1987), 158–59; Teaford, *Twentieth-Century American City,* 114; James Reichley, "Philadelphia Does It: The Battle for Penn Center," *Harper's* (February 1957); Dickinson, *This Is Greater Philadelphia,* 17–19; Horwood and Boyce, *Central Business District,* 57; Stern, Mellins, and Fishman, *New York, 1960,* 61–63, 167, 170; Spencer Klaw, "The New American Office," *Fortune* (September 1959); "The Great Manhattan Boom," *Time* (December 21, 1953); "The $2-Billion Building Boom," *Fortune* (February 1960); "Headquarters Town," *Fortune* (February 1960); Louis Schlivek, *Man in Metropolis* (Garden City, N.Y.: Doubleday, 1965), 93–94.

80. Walker Evans, "'Downtown': A Last Look Backward," *Fortune* (October 1956); Susan Lyman, *The Face of New York* (1954; reprint, New York: Crown, 1964).

81. "The New City," *Fortune* (February 1960); *Architectural Forum* (1960), cited in Stern, Mellins, and Fishman, *New York, 1960,* 175.

82. Editors of *Architectural Forum, Building, U.S.A.* (New York: McGraw-Hill, 1957), 83; see also Klaw, "The New American Office"; "Architecture's New Technology," *Fortune* (March 1956); Robin Boyd, "The Counter-Revolution in Architecture," *Harper's* (September 1959); Paul Goldberger, *The Skyscraper* (New York: Alfred A. Knopf, 1981), 103–13.

83. Lewis Mumford, *The Highway and the City* (New York: Harvest, 1963), 174.

84. Walker Evans, cited in Anthony Lane, "The Eye of the Land: How Walker Evans Reinvented American Photography," *New Yorker* (March 13, 2000), 91.

85. Cited in "Architecture's New Technology."

86. "Architecture's New Technology"; "The New City."

87. Stern, Mellins, and Fishman, *New York, 1960,* 345–46; "Architecture's New Technology."

88. Ada Louise Huxtable, cited in Stern, Mellins, and Fishman, *New York, 1960,* 330.

89. John Updike, *Assorted Prose* (New York: Alfred A. Knopf, 1965), 107.

90. Marya Mannes, *The New York I Know* (New York: J. B. Lippincott, 1961), 152–53.

91. Editors of *Architectural Forum, Building, U.S.A.,* 83, 68; Edward Durell Stone, "The Case Against the Tailfin Age," *New York Times Magazine* (October 18, 1959), 28.

92. Jane Krieger, "Office Beautiful in Big Business," *New York Times Magazine* (May 24, 1959); Klaw, "The New American Office"; Serge Guilbaut, *How New York Stole the Idea of Modern Art* (Chicago: University of Chicago Press, 1983), 89–90; Brooks, *The Great Leap,* 64.

93. Clement Greenberg, "After Abstract Expressionism," in Henry Geldzahler, ed., *New York Painting and Sculpture, 1940–1970* (New York: E. P. Dutton, 1969), 369; William Baziotes, cited in Irving Sandler, *The Triumph of American Painting: A History of Abstract Expressionism* (New York: Praeger, 1970), 93; Robert Motherwell, cited in Sandler, *Triumph of American Painting,* 96; "The Corporate Splurge in Abstract Art," *Fortune* (April 1960); "The Modern Art of Business," *Fortune* (March 1955).

94. Guilbaut, *How New York Stole,* 116, 142–43; Robert Hughes, *The Shock of the New* (New York: Alfred A. Knopf, 1991), 259; Barnett Newman, cited in Sandler, *Triumph of American Painting,* 149.

95. Clement Greenberg, *Art and Culture* (Boston: Beacon Press, 1961), 217, 137.

96. On modern art, see Philip Larkin, *All What Jazz?* (New York: Farrar, Straus and Giroux, 1985), 15–29.

97. Leo Steinberg, *Other Criteria* (New York: Oxford University Press, 1972), 15.

98. Earl Riley, cited in Abbott, *New Urban America,* 121; Cabell Phillips, "Exit the Boss, Enter the Leader," *New York Times Magazine* (April 15, 1956); Edward Banfield and James Q. Wilson, *City Politics* (Cambridge: Harvard University Press, 1967), 116–29; Edward Flynn, *You're the Boss* (New York: Viking, 1947), ix–x.

99. Lorant, *Pittsburgh,* 378; John Kay Adams, "Reforming Chicago," *Harper's* (June 1958); Len O'Connor, *Clout: Mayor Daley and His City* (Chicago: Regnery, 1975); "Clouter with Conscience," *Time* (March 15, 1963); Allan Talbot, *The Mayor's Game: Richard Lee of New Haven and the Politics of Change* (New York: Praeger, 1970), 46–58.

100. Banfield and Wilson, *City Politics,* 105–10; "Revolt in Philadelphia," *Saturday Evening Post* (November 8, 1952; November 15, 1952; and November 22, 1952); Aaron Levine, "Philadelphia Story: A New Look," *New York Times Magazine* (July 14, 1957); Joe Alex Morris, "How to Rescue a City," *Saturday Evening Post* (August 18, 1956); Robert E. Cantwell, "St. Louis Snaps Out of It," *Fortune* (July 1956); Avis Carlson, "St. Louis Wakes Itself Up," *Harper's* (March 1956); The Mayor's Wife, "My Husband Was Elected," *Atlantic Monthly* (November 1953); William Hessler, "Cincinnati: The City That Licked Corruption," *Harper's* (November 1953); William Hessler, "The Reform That Reformed Itself," *Reporter* (June 13, 1957); Abbott, *New Urban America,* 128–29, 130–31, 141–43; Edward Banfield, *Big City Politics* (New York: Random House, 1965), 11–12, 52–63; E. W. Kenworthy, "The Emergence of Mayor Wagner," *New York Times Magazine* (August 14, 1955); Talbot, *Mayor's Game;* Abbott, *New Urban America,* 248–49; editors of *Fortune, The Exploding Metropolis,* 81–109; Phillips, "Exit the Boss."

101. *Life* (June 6, 1955; and August 31, 1953); see also Leo Egan, "How New Yorkers Pick Their Mayors," *New York Times Magazine* (October 18, 1953).

102. Flynn, *You're the Boss,* 222–35; see also Robert Caro, *The Power Broker: Robert Moses and the Fall of New York* (New York: Alfred A. Knopf, 1974), 850–94.

103. Allen Raymond, "News Rhode Island Can't Get," *Reporter* (October 31, 1951); Ralph Martin, *The Bosses* (New York: G. P. Putnam, 1964), 167–208, 115–66; Richard Wallace, "The Twilight of Ed Crump," *Reporter* (August 1, 1950); David L. Cohn, "Sing No Blues for Memphis," *New York Times Magazine* (September 4, 1955).

104. Norman Thomas, "The City O'Dwyer Left Behind," *Reporter* (November 7, 1950); Caro, *Power Broker,* 756, 786–87; Raskin, "How the Docks Shape Up."

105. Caro, *Power Broker,* 787–88; see also *Life* (June 6, 1955); Banfield and Wilson, *City Politics,* 332–33.

106. Cited in Dan Wakefield, "Greenwich Village Challenges Tammany," *Commentary* (October 1959), 309; see also David Hurwood, "Grass-Roots Politics in Manhattan," *Atlantic Monthly* (October 1960).

107. Seymour Freedgood, "The Vacuum at City Hall," *Fortune* (February 1960); Robert Bendiner, "The Ghost of LaGuardia versus the Shadow of Dewey," *Reporter* (November 10, 1953); Robert Bendiner, "De Sapio's Big Moment; or, The Rout of the Innocents," *Reporter* (October 16, 1958); Nathan Glazer and Daniel Moynihan, *Beyond the Melting Pot* (1963; reprint, Cambridge: MIT Press, 1968), 5.

108. Hollis Alpert, "Philadelphia: Plans and Pigeons," *Partisan Review* (September–October 1950), 699.

109. "Revolt in Philadelphia"; Reichley, "Philadelphia Does It"; for a similar story in St. Louis, see R. E. Cantwell, "St. Louis"; J. A. Morris, "How to Rescue"; for Los Angeles, see Banfield, *Big City Politics,* 80–88.

110. Quote from Dickinson, *This Is Greater Philadelphia,* 26; Hannah Lees, "The Philadelphia Election: Crusaders and Machines," *Reporter* (October 20, 1955); Reichley, "Philadelphia Does It"; on GIs as reformers, see also Whalen, "Two Generals Patrol"; Schulberg, "How One Pier"; R. Martin, *The Bosses,* 156–57.

111. Lees, "Philadelphia Election"; Robert Bendiner, "New Trainer for the Donkey: Philadelphia's Finnegan," *Reporter* (October 4, 1956); James Reichley, "Dilworth's Dilemma," *Reporter* (October 29, 1959).

112. O'Connor, *Clout,* 54–55.

113. Quote from J. B. Martin, "How Corrupt Is Chicago?"; see also O'Connor, *Clout,* 84–86.

114. Joseph Bell, "Merriam of Chicago: Politician Without a Party," *Harper's* (November 1954), 56.

115. Adams, "Reforming Chicago"; Alan Ehrenhalt, *The Lost City* (New York: Basic Books, 1995), 46–54; "Clouter with Conscience"; Banfield, *Political Influence,* 244–49; Mayer and Wade, *Growth of a Metropolis,* 376, 377; Cohen and Taylor, *American Pharaoh,* 166–71.

116. "Clouter with Conscience"; Banfield, *Political Influence,* 16, 237–39; Ehrenhalt, *The Lost City,* 46–54; O'Connor, *Clout,* 131; Cohen and Taylor, *American Pharaoh,* 155–63.

117. Adams, "Reforming Chicago"; Bell, "Merriam of Chicago"; Ehrenhalt, *The Lost City,* 50–54.

118. John Madigan, "The Durable Mr. Dawson of Cook County, Illinois," *Reporter* (August 9, 1956); Banfield, *Big City Politics,* 107–18; James Q. Wilson, *Negro Politics* (New York: Free Press, 1965), 23–24; Paddy Bauler, cited in Ehrenhalt, *The Lost City,* 44.

119. "How One Big City Is Fighting for a Comeback," *U.S. News and World Report* (July 19, 1957); Banfield, *Big City Politics,* 52–54; Banfield and Wilson, *City Politics,* 116–17; R. Martin, *The Bosses,* 209–61; Harold Kaplan, *Urban Renewal Politics: Slum Clearance in Newark* (New York: Columbia University Press, 1963), 165, 166–83.

120. Abbott, *New Urban America,* 249–52, 124–43; Banfield, *Big City Politics,* 80–88.

121. Alpert, "Philadelphia," 699; Banfield, *Big City Politics,* 44, 22–24, 45–63, 80–88; Bruce Bliven, "Politics and TV," *Harper's* (November 1952).

122. R. Martin, *The Bosses,* 209–61.

123. Quoted in Burton, "Downtown Isn't Doomed!" 99; see also William Zeckendorf, "New Cities for Old," *Atlantic Monthly* (November 1951); "What's Happening to U.S. Cities?"; a series titled "What Next for Our American Cities?" *American City* (December 1949; January 1950; and February 1950); "Coordinate Transportation for Metropolitan Communities," *American City* (April 1952); Martin Anderson, *The Federal Bulldozer: A Critical Analysis of Urban Renewal, 1949–1962* (Cambridge: MIT Press, 1964).

124. Cited in Abbott, *New Urban America,* 249.

125. About St. Louis, see R. E. Cantwell, "St. Louis"; Carlson, "St. Louis Wakes"; J. A. Morris, "How to Rescue"; about Philadelphia, see "Revolt in

Philadelphia"; "Philadelphia Does It"; Alpert, "Philadelphia"; Dickinson, *This Is Greater Philadelphia*, 1; on New Haven, see Talbot, *Mayor's Game*, 116-35; Edward Logue, "Urban Ruin—or Urban Renewal," *New York Times Magazine* (November 9, 1958); about Pittsburgh, see Lorant, *Pittsburgh*, 374-75; "Pittsburgh Rebuilds"; Schriftgiesser, "The Pittsburgh Story"; Frank Hawkins, "Lawrence of Pittsburgh: Boss of the Mellon Patch," *Harper's* (August 1956); "Futurity Stake: Pittsburgh," *Vogue* (February 1, 1960); about Columbus, see "The Cities of America: Columbus, Ohio," *Saturday Evening Post* (May 3, 1952); about New York City, see Robert Moses, "New York *Has* a Future," *New York Times Magazine* (January 30, 1955); Stern, Mellins, and Fishman, *New York, 1960*; about Detroit, see "How One Big City Is Fighting"; Robert Mowitz and Deil Wright, *Profile of a Metropolis: A Case Book* (Detroit: Wayne State University Press, 1962), 11-79, 104-6, 141-68; about Newark, see Kaplan, *Urban Renewal Politics*, 93-113; on downtown highways in general, see Horwood and Boyce, *Central Business District*.

126. Quote from "Build Expressways Through Slum Areas," *American City* (November 1951); see also "Downtown Worcester Due for Transformation," *American City* (April 1952); Hal Burton, *The City Fights Back* (New York: Citadel Press, 1954); Talbot, *Mayor's Game*, 116-35; Raymond Mohl, "Race and Space in Miami," in Arnold Hirsch and Raymond Mohl, eds., *Urban Policy in Twentieth Century America* (New Brunswick, N.J.: Rutgers University Press, 1993).

127. See, for example, Mowitz and Wright, *Profile of a Metropolis*, 81-139; Chester Hartman, *The Transformation of San Francisco* (Totowa, N.J.: Rowman and Allenheld, 1984); Mohl, "Race and Space"; Talbot, *Mayor's Game*, 138-39.

128. Lorant, *Pittsburgh*, 373.

129. *New York Times* (October 30, 1963), cited in Elliot Willensky and Norval White, *AIA Guide to New York City* (New York: Harcourt, Brace, 1988), 869.

130. First quote from Moses, "New York *Has* a Future," 22; second quote from Lois Balcom, "The Best Hope for Our Big Cities," *Reporter* (October 3, 1957), 20; see also Hannah Lees, "Making Our Cities Fit to Live In," *Reporter* (February 21, 1957); Horwood and Boyce, *Central Business District*, 116, 126; Jane Jacobs, *The Death and Life of Great American Cities* (New York: Vintage, 1989), 154-71.

131. On redevelopment, see "Public Housing: An Important Local Issue," *American City* (January 1951); C. F. Palmer, "To Wipe Out the Slums," *New York Times Magazine* (December 16, 1956); "St. Louis Rebuilds," *American City* (August 1950); "Philadelphia Embarks on Vast Improvement Program," *American City* (May 1950); "Rebirth of the Cities," *Time* (December 5, 1955); Leo Adde, *Nine Cities: The Anatomy of Downtown Renewal* (Washington, D.C.: Urban Land Institute, 1969); Carlson, "St Louis Wakes"; Abbott, *New Urban America*, 146-69; Colean, *Renewing Our Cities*, 54, 131; Kaplan, *Urban Renewal Politics*, 26-27; Talbot, *Mayor's Game*, 116-35.

132. "What Next for Our American Cities?" (February 1950).

133. J. Jacobs, *Death and Life*, 165; see also Mannes, *New York I Know*; editors of *Fortune*, *The Exploding Metropolis*.

Chapter 3. The Neighborhoods

1. E. B. White, *Here Is New York* (1949; reprint, New York: Harper and Row, 1977), 26; Marya Mannes, *The New York I Know* (New York: J. B. Lippincott, 1961), 64–65; last quote from Bert Kemp, *EB* (Darien, Conn.: Paerdegat Park, 1998), 14.

2. White, *Here Is New York,* 24, 25–26.

3. Robert Caro, *The Power Broker* (New York: Vintage, 1975), 851–53.

4. Federal Writers' Project, *The WPA Guide to New York City* (1939; reprint, New York: Pantheon, 1982), 498; Alfred Kazin, *A Walker in the City* (1951; reprint, New York: Harcourt, Brace, 1979), 35; man quoted in Jonathan Rieder, *Canarsie: The Jews and Italians of Brooklyn Against Liberalism* (Cambridge: Harvard University Press, 1985), 25; see also Irving Shulman, *The Amboy Dukes* (1947; reprint, New York: Bantam, 1965), 3.

5. Margot Patterson Doss, *San Francisco at Your Feet* (New York: Grove Press, 1964), 104–5; see also Coro Foundation, *The District Handbook* (San Francisco: Coro Foundation, 1979), 215.

6. Alan Ehrenhalt, *The Lost City* (New York: Basic Books, 1995), 98–99; "Shopping After Dark," *Fortune* (March 1952).

7. Elliot Willensky, *When Brooklyn Was the World, 1920–1957* (New York: Harmony Books, 1986), 150; Doris Kearns Goodwin, *Wait till Next Year* (New York: Simon and Schuster, 1997), 66–68; Kemp, *EB,* 99.

8. Irving Shulman, *Cry Tough!* (New York: Avon, 1950), 5, 47, 52; quote about mafiosi cited in Rieder, *Canarsie,* 25.

9. Shulman, *Cry Tough!,* 161; see also Ehrenhalt, *The Lost City,* 50–54; Frank Hawkins, "Lawrence of Pittsburgh: Boss of the Mellon Patch," *Harper's* (August 1956), 58.

10. Willensky, *When Brooklyn Was,* 150–51.

11. Ibid., 153.

12. The stores listed above, and many others, were on a three-block strip of Polk Street in San Francisco in 1950. See Sonia Lehman Frisch, "Le rôle de la rue commerçante dans l'identité de quartier à San Francisco" (Ph.D. diss., Geography Department, University of Paris, Nanterre, 2002); quote from Susan Lyman, *The Face of New York* (1954; reprint, New York: Crown, 1964).

13. Mannes, *New York I Know,* 64–65; Goodwin, *Wait till Next Year,* 66–68.

14. Kemp, *EB,* 99–100.

15. Pete Hamill, *A Drinking Life: A Memoir* (Boston: Little, Brown, 1994), 24, 26; John Cheever, *The Brigadier and the Golf Widow* (New York: Harper and Row, 1964), 71.

16. Cited in Lee Rainwater et al., *Workingman's Wife* (New York: Oceana, 1959), 164, 163; see also Lyman, *Face of New York.*

17. Ehrenhalt, *The Lost City,* 104–8.

18. Donald Paneth, "I Cash Clothes!" *Commentary* (June 1950), 556, 558.

19. Mike Royko, *The Best of Royko* (Chicago: University of Chicago Press, 1999), 7–9; see also Donald Paneth, "McCaffrey's Bar and Grill: The Neighbor-

hood Club on the Corner," *Commentary* (July 1953); Kemp, *EB,* 148–51; Hamill, *A Drinking Life,* 61–62, 166–67; Mannes, *New York I Know,* 64–65.

20. Quote from John Kay Adams, "Reforming Chicago: Slow but Not Hopeless," *Harper's* (June 1958), 69; see also Len O'Connor, *Clout: Mayor Daley and His City* (Chicago: Regnery, 1975), 84–86, 119, 131; Edward Banfield, *Political Influence* (New York: Free Press, 1961), 16, 237–39; Ehrenhalt, *The Lost City,* 46–49; Edward Banfield, *Big City Politics* (New York: Random House, 1965), 107–18; Edward Banfield and James Q. Wilson, *City Politics* (Cambridge: Harvard University Press, 1967), 116–29; Edward Flynn, *You're the Boss* (New York: Viking, 1947), 222–35; Hamill, *A Drinking Life,* 61, 68. Paul Jacobs, *The State of the Unions* (New York: Atheneum, 1963), 27; Virgil Peterson, "The Chicago Police Scandal," *Atlantic Monthly* (October 1960).

21. Kemp, *EB,* 7, 151; see also Hamill, *A Drinking Life,* 30, 61–62, 166–67; Richard Yates, *Eleven Kinds of Loneliness* (Boston: Little, Brown, 1962).

22. Hamill, *A Drinking Life,* 60; Mike Royko, *Boss: Richard J. Daley of Chicago* (New York: Penguin, 1976), 31–32.

23. First quote from Ehrenhalt, *The Lost City,* 94–95; 80 percent statistic from Gerald Suttles, *The Social Order of the Slum* (Chicago: University of Chicago Press, 1971), 76–77; second quote from Kemp, *EB,* 16, 75; final quote from Shulman, *Cry Tough!,* 116.

24. Suttles, *Social Order,* 75–76; Hamill, *A Drinking Life,* 186.

25. Quote from Elizabeth Cromley, *Alone Together* (Ithaca, N.Y.: Cornell University Press, 1990), xiii. On women working for wages, see Edwin Goldfield, *Statistical Abstract of the United States, 1965* (Washington, D.C.: U.S. Government Printing Office, 1965), 226, 228; Paul Glick, *American Families* (New York: Wiley and Sons, 1957), 92–93; Robert O. Blood and Donald M. Wolfe, *Husbands and Wives: The Dynamics of Married Living* (New York: Free Press, 1960), 99.

26. Herbert Gans, *The Urban Villagers: Group and Class in the Life of Italian-Americans* (1962; reprint, New York: Free Press, 1982), 105–6; Rainwater et al., *Workingman's Wife,* 34–37, 44–49, 69, 103–5, 114–17; Ehrenhalt, *The Lost City,* 112–35; John McGreevy, *Parish Boundaries: The Catholic Encounter with Race in the Twentieth-Century Urban North* (Chicago: University of Chicago Press, 1996), 79, 80, 120–25; Arthur Shostak and William Gomberg, *Blue-Collar World* (Englewood Cliffs, N.J.: Prentice-Hall, 1964); Mirra Komarovsky, *Blue-Collar Marriage* (New York: Random House, 1962).

27. Ehrenhalt, *The Lost City,* 97; Gans, *Urban Villagers,* 14; Kemp, *EB,* 38–62, 80–91; Kazin, *Walker in the City,* 77–131.

28. Edgar Freidenberg, *Coming of Age in America* (New York: Random House, 1965).

29. Dwight Macdonald, "A Caste, a Culture, a Market," parts 1 and 2, *New Yorker* (November 22, 1958), 57; (November 29, 1958); James Coleman, *The Adolescent Society* (New York: Free Press, 1961), 11; last quote from Frederick Lewis Allen, "The Spirit of the Times," *Harper's* (July 1952); see also Robert Lindner, *Must You Conform?* (New York: Rinehart, 1956), 6; David Riesman, "Where Is the College Generation Headed?" *Atlantic Monthly* (April 1961), 39; Agnes Rogers,

"The Humble Female," *Harper's* (March 1950), 58; H. H. Remmers and D. H. Sadler, *The American Teenager* (New York: Bobbs-Merrill, 1957).

30. Macdonald, "A Caste, a Culture, a Market," parts 1 and 2; Eve Merriam, *Figleaf* (New York: J. B. Lippincott, 1960), 140–41; James Gilbert, *A Cycle of Outrage* (New York: Oxford University Press, 1986), 98, 13, 196, 204–5.

31. Arnold Shaw, *The Rockin' 50s* (New York: Hawthorn, 1974), 122–26, 66–72; Kemp, *EB*, 77; Hamill, *A Drinking Life*, 144; David Dachs, *Anything Goes: The World of Popular Music* (New York: Bobbs-Merrill, 1964), 41; Charles Gillett, *The Sound of the City* (New York: Pantheon, 1983), 207–8; James Miller, *Flowers in the Dustbin: The Rise of Rock and Roll, 1947–1977* (New York: Simon and Schuster, 1999), 145–46.

32. William Graebner, *Coming of Age in Buffalo* (Philadelphia: Temple University Press, 1992), 98, 27–28, 34–39.

33. Ibid., 27–28.

34. Shaw, *The Rockin' 50s*, 63, 66–72, Frank Zappa quote, 122; Graebner, *Coming of Age*, 29–32; Miller, *Flowers in the Dustbin*, 37, 72, 76, 77, 86, 89–93, 106, 107, 122–26, 163–64, 321; Gillett, *Sound of the City*, 3, 64.

35. Hubert Selby, *Last Exit to Brooklyn* (1957; reprint, New York: Grove Press, 1988), 11–12; Shulman, *The Amboy Dukes* and *Cry Tough!*; Evan Hunter, *The Blackboard Jungle* (New York: Simon and Schuster, 1954); Kemp, *EB*, 103–4.

36. Comment about the Gunners cited in Graebner, *Coming of Age*, 53; see also Willensky, *When Brooklyn Was*, 133; Shulman, *The Amboy Dukes* and *Cry Tough!*; Murray Schumach, "The Teen-Age Gang: Who and Why," *New York Times Magazine* (September 2, 1956); Will Chasen, "Teen-Age Gang from the Inside," *New York Times Magazine* (March 21, 1954); Virginia Held, "What Can We Do about 'J.D.?'" *Reporter* (August 20, 1959); Hamill, *A Drinking Life*, 96–97; *Saturday Evening Post* (January 8, 1955; January 15, 1955; January 22, 1955; January 29, 1955; February 5, 1955); *Life* (August 12, 1957; August 26, 1957; September 9, 1957); Gilbert, *A Cycle of Outrage*.

37. Cited in Macdonald, "A Caste, a Culture, a Market," part 1, 83; see also Benita Eisler, *Private Lives: Men and Women of the Fifties* (New York: Franklin Watts, 1986), 96–97; Graebner, *Coming of Age*, 22–23; Nora Sayre, *Previous Convictions: A Journey Through the 1950s* (New Brunswick, N.J.: Rutgers University Press, 1995), 112.

38. Harold Kaplan, *Urban Renewal Politics: Slum Clearance in Newark* (New York: Columbia University Press, 1963), 150; Patricia Cayo Sexton, *Spanish Harlem* (New York: Harper and Row, 1965); Dan Wakefield, *Island in the City: The World of Spanish Harlem* (Boston: Houghton Mifflin, 1959); Brian Godfrey, *Neighborhoods in Transition* (Berkeley: University of California Press, 1988), 131–64, 182–89, 121; William Barrett, "New Designs in Our Bohemia," *New York Times Magazine* (August 20, 1950); James T. Farrell, *Boarding House Blues* (New York: Paperback Library, 1961).

39. Kemp, *EB*, 14.

40. John Bodnar et al., *Lives of Their Own: Blacks, Italians, and Poles in Pittsburgh, 1900–1960* (Chicago: University of Illinois Press, 1982), 219–20, 229; see also McGreevy, *Parish Boundaries*, 83–84, 131–32.

41. Quote from Hannah Lees, "How Philadelphia Stopped a Race Riot," *Reporter* (June 2, 1955), 26; Bodnar et al., *Lives of Their Own,* 220.

42. Gerald Weales, "Small-Town Detroit," *Commentary* (September 1956), 219–26; Royko, *Boss,* 32; see also Norman Podhoretz, *Making It* (New York: Random House, 1967), 83.

43. Jacques Barzun, *God's Country and Mine* (Boston: Little, Brown, 1954), 10; Goldfield, *Statistical Abstract,* 92; Ben Wattenberg, *This U.S.A.: An Unexpected Family Portrait of 194,067,296 Americans Drawn from the Census* (Garden City, N.Y.: Doubleday, 1965), 359.

44. Gans, *Urban Villagers,* 33–34; Oscar Handlin, "We Need More Immigrants," *Atlantic Monthly* (May 1953); Robert Bremner and Gary Reichard, eds., *Reshaping America: Society and Institutions, 1945–1960* (Columbus: Ohio State University Press, 1982), 165; Dan Wakefield, "New York's Lower East Side Today," *Commentary* (June 1959); Morris Janowitz, *The Community Press in an Urban Setting* (Chicago: University of Chicago Press, 1967), xiv, 22–24, 61–77; Joshua Freeman, *Working Class New York* (New York: New Press, 2000), 25, 27.

45. Warner Bloomberg, "The State of the American Proletariat, 1955," *Commentary* (March 1955), 211; on conformity, see Lindner, *Must You Conform?;* Herbert Marcuse, *One Dimensional Man* (Boston: Beacon Press, 1964); C. Wright Mills, *White Collar* (New York: Oxford University Press, 1953); Irving Howe, "This Age of Conformity," *Partisan Review* (January–February, 1954); William Whyte, *The Organization Man* (New York: Anchor, 1957); Alan Valentine, *The Age of Conformity* (Chicago: Regnery, 1954); William S. White, "'Consensus American': A Portrait," *New York Times Magazine* (November 25, 1956); Daniel Seligman, "The New Masses," *Fortune* (May 1959).

46. Kazin, *Walker in the City,* 11, *passim.*

47. Bloomberg "American Proletariat," 211.

48. Quote from Lyman, *The Face of New York,* n.p.; Will Herberg, *Protestant-Catholic-Jew: An Essay in American Religious Sociology* (Garden City, N.Y.: Doubleday, 1956); McGreevy, *Parish Boundaries,* 79–82; Seligman, "The New Masses," 258.

49. Philip Roth, interview by Terry Gross, *Fresh Air,* National Public Radio (May 8, 2000); Saul Bellow, *The Adventures of Augie March* (1949; reprint, New York: Viking, 1960), 3.

50. First quote by George Meany, in Stephen Ambrose, *Eisenhower: Soldier and President* (New York: Simon and Schuster, 1990), 386; second in George Meany, "What Labor Means by 'More,'" *Fortune* (March 1955), 92.

51. Raymond Vernon, *The Changing Economic Function of the Central City* (New York: New York Metropolitan Region Study, 1959), 32–33, 50–51, 70–73, 74–79; Robert H. Zieger, *American Workers, American Unions* (Baltimore: Johns Hopkins University Press, 1994), 138; Harold Mayer and Richard Wade, *Chicago: Growth of a Metropolis* (Chicago: University of Chicago Press, 1969), 426; William B. Dickinson, *This Is Greater Philadelphia* (Philadelphia: Bulletin Company, 1954), 5; Peter Muller, *Contemporary Suburban America* (Englewood Cliffs, N.J.: Prentice-Hall, 1981), 135, 148; "Is Downtown Doomed?" *Saturday Evening Post* (January 1, 1960); E. Willard Miller, *A Geography of Manufacturing* (Englewood Cliffs, N.J.: Prentice-Hall, 1962); Freeman, *Working Class New York,* 3–200.

52. Dickinson, *This Is Greater Philadelphia*, 4; E. W. Miller, *A Geography of Manufacturing*, 40; Thomas Sugrue, *The Origins of the Urban Crisis: Race and Inequality in Postwar Detroit* (Princeton, N.J.: Princeton University Press, 1996), 125–26.

53. Peter Cohen, "Transformation in an Industrial Landscape: San Francisco's Northeast Mission" (master's thesis, Geography Department, San Francisco State University, 1998), 93–97; see also Freeman, *Working Class New York*, 143.

54. William O'Neill, *American High: The Years of Confidence, 1945–1960* (New York: Free Press, 1986), 90–91; Daniel Bell, "Beyond the 'Annual Wage,'" *Fortune* (May 1955); Daniel Bell, "No Boom for the Unions," *Fortune* (June 1956); Samuel Bowles, David Gordon, and Thomas Weisskopf, *After the Waste Land* (New York: Sharpe, 1990), 68; Zieger, *American Workers, American Unions*, 75–139; "Labor," *Fortune* (January 1955).

55. George Gallup, *The Gallup Poll: Public Opinion, 1935–1971* (New York: Random House, 1972), 2: 1184, 1469, 1484, 1516; for coverage of the hearings, see *Life* (April 9, 1956; March 1, 1957; March 11, 1957; March 25, 1957; September 9, 1957; October 14, 1957; December 9, 1957); on Hoffa and the teamsters, see Bernard Nossiter, "The Teamsters: Corrupt Policemen of an Unruly Industry," *Harper's* (May 1959); Clark Mellenhoff, "The Teamsters Defy the Government," *Atlantic Monthly* (November 1958); Mary Heaton Vorse, "The Pirate's Nest of New York," *Harper's* (April 1952); Sumner Slichter, "New Goals for the Unions," *Atlantic Monthly* (December 1958); Edward Chamberlin, "Can Union Power Be Curbed?" *Atlantic Monthly* (June 1959); Sidney Lens, "Dave Beck's Teamsters: Sour Note in Labor Harmony," *Harper's* (February 1956); Jacobs, *State of the Unions*, 5–87, 93–100; A. H. Raskin, "Reuther vs. Hoffa: A Key Struggle," *New York Times Magazine* (September 22, 1957); A. H. Raskin, "Labor's House Three Years After," *New York Times Magazine* (November 30, 1958).

56. Daniel Bell, "The Language of Labor," *Fortune* (September 1951), 86; A. H. Raskin, "Union Leader—and Big Business Man," *New York Times Magazine* (November 15, 1953); *Saturday Evening Post* (April 18, 1953).

57. Wallace Markfield, "Seventh Avenue: Boss and Worker," *Commentary* (March 1950), 268; Jacobs, *State of the Unions*, 112–36; Warner Bloomberg, "The Vote-Getter of Lake County, Indiana," *Reporter* (October 28, 1952); Warner Bloomberg, "How They Took the Bad News in Gary, Indiana," *Reporter* (December 9, 1952); Freeman, *Working Class New York*, 55–71.

58. Quote from Clancy Sigal, *Going Away: A Report, A Memoir* (Boston: Houghton Mifflin, 1961), 73; see also Warner Bloomberg, "Report from Gary: 'It Wasn't So Bad,'" *Reporter* (September 2, 1952); Warner Bloomberg, "Time Off in Gary," *Reporter* (June 24, 1952); A. H. Raskin, "Labor: A New 'Era of Bad Feeling'?" *New York Times Magazine* (July 5, 1959); A. H. Raskin, "Deep Shadow over Our Factories," *New York Times Magazine* (November 29, 1959); Freeman, *Working Class New York*, 55–104.

59. Warner Bloomberg, "Five Hot Days in Gary," *Reporter* (August 11, 1955), 38; A. H. Raskin, "The Outlook for Labor Under Eisenhower," *Commentary* (April 1953), 372; Meany, "What Labor Means," 92; see also A. H. Raskin, "Mood of

United Labor After a Year," *New York Times Magazine* (December 2, 1956); Harvey Swados, *On the Line* (Boston: Little, Brown, 1957).

60. Oscar Handlin, "Payroll Prosperity: Will American Labor Go Conservative?" *Atlantic Monthly* (February 1953); see also Shostak and Gomberg, *Blue-Collar World*, 98; A. H. Rosen, "Labor's Time of Trouble," *Commentary* (August 1959); Joe Miller, "Dave Beck Comes out of the West," *Reporter* (December 8, 1953); Murray Kempton, "Labor: The Alliance on the Plateau," *Reporter* (June 30, 1955); Raskin, "Mood of United Labor"; Swados, *On the Line*, 71–102; Freeman, *Working Class New York*, 166.

61. Marya Mannes, "Labor Day on Fifth Avenue," *Reporter* (October 1, 1959); Zieger, *American Workers, American Unions*, 120–21, 158; Raskin, "Outlook for Labor," 366; Gallup, *Gallup Poll*, 2: 1137, 1200; see also Bloomberg, "Bad News in Gary, Indiana."

62. Rainwater et al., *Workingman's Wife*, 17; "The Rich Middle-Income Class," *Fortune* (May 1954), 95; Gilbert Burck and Sanford Parker, "The Changing American Market," *Fortune* (August 1953); Handlin, "Payroll Prosperity"; Daniel Bell, "Labor's Coming of Middle Age," *Fortune* (October 1951); Zieger, *American Workers, American Unions*, 120–61.

63. Geraldine Fregoso, "Growing Up in the Mission," *San Francisco Sunday Examiner and Chronicle* (April 21, 1974); Graebner, *Coming of Age*, 74–75.

64. Albert Votaw, "The Hillbillies Invade Chicago," *Harper's* (February 1958), 65; see also editors of *Fortune*, *The Exploding Metropolis* (Garden City, N.Y.: Doubleday, 1958), 114; Weales, "Small-Town Detroit"; Harriette Arnow, *The Dollmaker* (1954; reprint, New York: Avon, 1972); James Maxwell, "Down from the Hills and into the Slums," *Reporter* (December 13, 1956).

65. Oscar Handlin, *The Newcomers* (New York: Anchor, 1962), 52–58; Kaplan, *Urban Renewal Politics*, 150; Federal Writers' Project, *WPA Guide*, 265–68; Sexton, *Spanish Harlem*; Wakefield, *Island in the City*; Edward Rivera, *Family Installments* (New York: William Morrow, 1982); Oscar Lewis, *La Vida: A Puerto Rican Family in the Culture of Poverty — San Juan and New York* (New York: Random House, 1966).

66. Quote from Mannes, *New York I Know*, 18–22; Julius Horwitz, *The Inhabitants* (New York: Signet, 1960); Louis Schlivek, *Man in Metropolis* (Garden City, N.Y.: Doubleday, 1965), 213–14; Anzia Yezierska, "The Lower Depths of Upper Broadway," *Reporter* (January 19, 1954); Robert Stern, Thomas Mellins, and David Fishman, *New York, 1960* (New York: Monacelli Press, 1995), 663–64.

67. Jon Teaford, *The Twentieth-Century American City* (Baltimore: Johns Hopkins University Press, 1986), 115–16; Stephen Thernstrom, *The Other Bostonians: Poverty and Progress in the American Metropolis, 1870–1970* (Cambridge: Harvard University Press, 1973), 179; Kaplan, *Urban Renewal Politics*, 149; Goldfield, *Statistical Abstract*, 19–20.

68. Arnold Hirsch, *Making the Second Ghetto: Race and Housing in Chicago, 1940–1960* (Cambridge: Cambridge University Press, 1988), 28.

69. Weales, "Small-Town Detroit," 219–26; see also Sugrue, *Urban Crisis*, 209–58; Banfield, *Big City Politics*, 52.

70. Teaford, *Twentieth-Century American City,* 115; Graebner, *Coming of Age,* 79–85; Stern, Mellins, and Fishman, *New York, 1960,* 917, 922; Alter Landesman, *Brownsville: The Birth, Development, and Passing of a Jewish Community in New York* (New York: Bloch, 1969), 371–72; Kazin, *Walker in the City,* 6–12; Rieder, *Canarsie;* Steven Gregory, *Black Corona: Race and the Politics of Place in an Urban Community* (Princeton, N.J.: Princeton University Press, 1998), 60, 61–63; McGreevy, *Parish Boundaries,* 102; Studs Terkel, *Division Street, America* (New York: Pantheon, 1967), 127–29; Hirsch, *Making the Second Ghetto,* 68–99; Kaplan, *Urban Renewal Politics,* 150; Albert Broussard, *Black San Francisco: The Struggle for Racial Equality in the West, 1900–1954* (Lawrence: University of Kansas Press, 1993), 133–245; Bodnar et al., *Lives of Their Own,* 229–31; Raymond Mohl, "Race and Space in Miami," in Arnold Hirsch and Raymond Mohl, eds., *Urban Policy in Twentieth Century America* (New Brunswick, N.J.: Rutgers University Press, 1993), 102–4, 129–34.

71. Dickinson, *This Is Greater Philadelphia,* 33–35; see also McGreevy, *Parish Boundaries,* 103–4.

72. Dickinson, *This Is Greater Philadelphia,* 34.

73. McGreevy, *Parish Boundaries,* 91–92, 107.

74. Lees, "Philadelphia Stopped a Race Riot"; McGreevy, *Parish Boundaries,* 91–101, 106–7; St. Clair Drake and Horace Cayton, *Black Metropolis* (1945; reprint, Chicago: University of Chicago Press, 1993), 817, 818; Hirsch, *Making the Second Ghetto,* 68–99.

75. *Chicago Defender* (July 21, 1951); *Life* (July 23, 1951); see also *Chicago Defender* (October 6, 1951); Hirsch, *Making the Second Ghetto,* 53–55.

76. Cited in Gregory, *Black Corona,* 60.

77. Drake and Cayton, *Black Metropolis,* 793.

78. First quote in Weales, "Small-Town Detroit," 219; second quote cited in Sugrue, *Urban Crisis,* 36.

79. *Ebony* magazine and actress on Sugar Hill, cited in Jervis Anderson, *This Was Harlem* (New York: Farrar, Straus and Giroux, 1982), 341–42, 344; see also Lawrence Graham, *Our Kind of People: Inside America's Black Upper Class* (New York: HarperCollins, 1999), 260.

80. Drake and Cayton, *Black Metropolis,* 660, 820–22; Graham, *Our Kind of People,* 202–4, 304, 312–13; Stern, Mellins, and Fishman, *New York, 1960,* 917; Gregory, *Black Corona,* 63–66; Kaplan, *Urban Renewal Politics,* 152–53.

81. Drake and Cayton, *Black Metropolis,* 600–57, 796–98; Elliot Liebow, *Talley's Corner: A Study of Negro Streetcorner Men* (Boston: Little, Brown, 1967), 3–27; James Conant, *Slums and Suburbs* (New York: McGraw-Hill, 1961), 24.

82. Harrison Salisbury, *The Shook-Up Generation* (New York: Harper, 1958), 11; see also "The Cougars: Life with a Brooklyn Gang," *Harper's* (November 1954).

83. Langston Hughes, *The Best of Simple* (New York: Hill and Wang, 1961), 21.

84. "Look Back to the Upper Fillmore," http://www.e-media.com/fillmore/museum/mintun01.html; Elizabeth Pepin, "Swing the Fillmore," *San Francisco Bay Guardian* (February 25, 1998).

85. *Sun Reporter* (March 1, 1952; and March 8, 1952); Broussard, *Black San Francisco,* 221, 232–33.

86. Pepin, "Swing the Fillmore."

87. Jimmy Witherspoon, cited in Johnny Otis, *Upside Your Head! Rhythm and Blues on Central Avenue* (Hanover, N.H.: University Press of New England, 1993), 4; Ted Gioia, *West Coast Jazz* (Berkeley: University of California Press, 1998), 3–15; Art Pepper and Laurie Pepper, *Straight Life* (London: Schirmer, 1979), 41–42, 45–46, 113–14; Michael Solot, "It's Really Gone, Man," *Chicago Reader* (May 7, 1993).

88. Howard Reich, "Where Jazz Came to Stay," *Chicago Tribune* (January 9, 1997).

89. David Rosenthal, *Hard Bop: Jazz and Black Music, 1955–1965* (New York: Oxford University Press, 1992), 4–5, 8, 25, 62–69, 171.

90. Solot, "It's Really Gone, Man," 10.

91. I base this on having read San Francisco's *Sun Reporter* 1950s-era issues and several years of the *Pittsburgh Courier,* the *N.Y. Age Defender,* and the *Chicago Defender;* see also James Baldwin, *Notes of a Native Son* (1955; reprint, Boston: Beacon Press, 1990), 60–65; Drake and Cayton, *Black Metropolis,* 398–412, 793.

92. *Sun Reporter* (March 9, 1957); *Chicago Defender* (January 5, 1957).

93. *Sun Reporter* (February 16, 1952; February 23, 1952; March 1, 1952; and May 3, 1952); see also Broussard, *Black San Francisco,* 180–92, 221–38; Graham, *Our Kind of People, passim;* Drake and Cayton, *Black Metropolis,* 798–802.

94. Graham, *Our Kind of People,* 305.

95. Cited in Gregory, *Black Corona,* 65.

96. E. Franklin Frazier, *Black Bourgeoisie* (1957; reprint, London: Collier, 1970), 50–55, 77, 92–97, 140–45; Drake and Cayton, *Black Metropolis,* 794–95, 803–4; see also Broussard, *Black San Francisco,* 186; Robert Kinzer and Edward Sagarin, *The Negro in American Business* (New York: Greenberg, 1950), 2–93, 108, 122; Emmet John Hughes, "The Negro's New Economic Life," *Fortune* (September 1956), 254; Nathan Glazer and Daniel Moynihan, *Beyond the Melting Pot* (1963; reprint, Cambridge: MIT Press, 1968), 53.

97. *Sun Reporter* (February 16, 1952).

98. Baldwin, *Native Son,* 68.

99. Glazer and Moynihan, *Beyond the Melting Pot,* 30–31, 71–73.

100. Edward Lewis Wallant, *The Pawnbroker* (New York: MacFadden, 1961), 15–16.

101. John Madigan, "The Durable Mr. Dawson of Cook County, Illinois," *Reporter* (August 9, 1956); Ralph Martin, *The Bosses* (New York: G. P. Putnam, 1964), 263–94; Banfield, *Big City Politics,* 117–18; Handlin, *The Newcomers,* 112; Frazier, *Black Bourgeoisie,* 150–61; Drake and Cayton, *Black Metropolis,* 794; James Q. Wilson, "How the Northern Negro Uses His Vote," *Reporter* (March 31, 1960).

102. James Q. Wilson, *Negro Politics* (New York: Free Press, 1965), 177–213; Drake and Cayton, *Black Metropolis,* 802; Kaplan, *Urban Renewal Politics,* 153–56; Martin, *The Bosses,* 263–94; Hirsch, *Making the Second Ghetto,* 245–53.

103. Cited in Anderson, *This Was Harlem,* 349.

104. Russell Baker, *The Good Times* (New York: William Morrow, 1989), 77; E. J. Hughes, "Negro's New Economic Life," 251; Douglas Cater, "Atlanta: Smart Politics and Good Race Relations," *Reporter* (July 11, 1957).

105. Ralph Bunche, cited in Irwin Ross, "Ralph Bunche, Mediator," *Reader's Digest* (April 1950), 146; Chester Bowles, "The Negro-Progress and Challenge," *New York Times Magazine* (February 7, 1954); *Chicago Defender* (October 6, 1951).

106. George Long, cited in Daniel Wolff, "There Always Was Pride," *Doubletake* (Winter 1999), 56; on Memphis, see Richard Wallace, "The Twilight of Ed Crump," *Reporter* (August 1, 1950); Graham, *Our Kind of People*, 277, 285–86; for other southern cities, see Carl Abbott, *The New Urban America: Growth and Politics in the Sunbelt Cities* (Chapel Hill: University of North Carolina Press, 1987), 250; Banfield, *Big City Politics*, 23–24.

107. *New York Amsterdam Times* (May 5, 1962); for similar stories about New Orleans, see *Chicago Defender* (January 12, 1957) and *Pittsburgh Courier* (January 7, 1956).

108. L. Hughes, *The Best of Simple*, 20–21.

109. *Pittsburgh Courier* (August 23, 1952).

110. Bodnar et al., *Lives of Their Own*, 249–52; Thernstrom, *Other Bostonians*, 197–219; Broussard, *Black San Francisco*, 241; Blood and Wolfe, *Husbands and Wives*, 99.

111. E. J. Hughes, "Negro's New Economic Life," 127–31, 254; Drake and Cayton, *Black Metropolis*, 807–16; Bowles, "Negro-Progress and Challenge"; Wattenberg, *This U.S.A.*, 279.

112. Drake and Cayton, *Black Metropolis*, 794.

113. First quote from *Newsweek* (July 27, 1959); second quote from Drake and Cayton, *Black Metropolis*, 796; see also *Chicago Defender* (March 19, 1955); Herbert Hill, "Labor Unions and the Negro," *Commentary* (December 1959); Jacobs, *State of the Unions*, 154–69; E. J. Hughes, "Negro's New Economic Life," 127–31, 256; Will Chasen, "American Labor Attacks Its Own Segregation Problem," *Reporter* (May 1, 1958); *Pittsburgh Courier* (January 7, 1956); Freeman, *Working Class New York*, 179–200; Thernstrom, *Other Bostonians*, 197–219; Broussard, *Black San Francisco*, 241, 214, 217.

114. Hirsch, *Making the Second Ghetto*, 124–31, 226; Robert Mowitz and Deil Wright, *Profile of a Metropolis: A Case Book* (Detroit: Wayne State University, 1962), 11–79; "Is Downtown Doomed?"; *Newsweek* (July 27, 1959); Salisbury, *The Shook-Up Generation*, 80–81.

115. Thomas Farrell, "Object Lesson in Race Relations," *New York Times Magazine* (February 12, 1950); Kaplan, *Urban Renewal Politics*, 147–63; Drake and Cayton, *Black Metropolis*, 660–61; Mary Wilson, cited in Wini Breines, *Young, White, and Miserable: Growing Up Female in the Fifties* (Boston: Beacon Press, 1992), 206.

116. George Sternlieb, *The Tenement Landlord* (New Brunswick, N.J.: Urban Studies Center, 1966), 36; *Chicago Defender* (March 19, 1955); Sugrue, *Urban Crisis*, 43, 194–97; Hannah Lees, "Negro Neighbors," *Atlantic Monthly* (January 1956); Hirsch, *Making the Second Ghetto*, 28–39; E. J. Hughes, "Negro's New Economic Life, 258; Liebow, *Talley's Corner*, 41; Handlin, *The Newcomers*, 90; Broussard, *Black San Francisco*, 221–38; Oscar Steiner, "Slums Are a Luxury We Cannot Afford," *Reporter* (November 14, 1957); Drake and Cayton, *Black Metropolis*, 816–25; McGreevy, *Parish Boundaries*, 104.

117. *Chicago Defender* (January 26, 1957); E. J. Hughes, "Negro's New Economic Life," 127; see also Bowles, "Negro-Progress and Challenge"; E. W. Kenworthy, "Taps for Jim Crow in the Services," *New York Times Magazine* (June 11, 1950).

118. J. Ronald Oakley, *God's Country: America in the Fifties* (New York: W. W. Norton, 1986), 38–39.

119. *Pittsburgh Courier* (August 30, 1952); Conant, *Slums and Suburbs*, 18; Broussard, *Black San Francisco*, 206, 209, 235–36; see also E. J. Hughes, "Negro's New Economic Life," 127; Bowles, "Negro-Progress and Challenge"; Kenworthy, "Taps for Jim Crow."

120. Drake and Cayton, *Black Metropolis*, 794; Dawson quoted in Madigan, "Durable Mr. Dawson," 41.

121. Broussard, *Black San Francisco*, 231.

122. On the NAACP, see *Chicago Defender* (January 19, 1957); on Jim Crow train, see *Pittsburgh Courier* (January 7, 1956); see also Drake and Cayton, *Black Metropolis*, 793–825; Nelson George, *The Death of Rhythm and Blues* (New York: Pantheon, 1988), 51; Lees, "Philadelphia Stopped a Race Riot."

123. David Halberstam, "'A Good City Gone Ugly,'" *Reporter* (March 31, 1960), 19.

124. Eric Hobsbawm, *The Jazz Scene* (New York: Pantheon, 1993), xxvi; Rosenthal, *Hard Bop*, 171; Janowitz, *Community Press*, xii; Drake and Cayton, *Black Metropolis*, xlvii–lii, 826–36; Kenneth Clark, *Dark Ghetto* (New York: Harper, 1965).

Chapter 4. The Suburbs

1. *Time* (July 3, 1950); John McPartland, *No Down Payment* (New York: Simon and Schuster, 1957), 18; "The Lush New Suburban Market," *Fortune* (November 1953), 128.

2. Kenneth Jackson, *Crabgrass Frontier* (New York: Oxford University Press, 1985), 232; Robert Stern, Thomas Mellins, and David Fishman, *New York, 1960* (New York: Monacelli Press, 1995), 14.

3. Man quoted in Louis Schlivek, *Man in Metropolis* (Garden City, N.Y.: Doubleday, 1965), 30; see also Robert Moses, "Build and Be Damned," *Atlantic Monthly* (December 1950), 40; Jackson, *Crabgrass Frontier*, 206.

4. McPartland, *No Down Payment*, 24; see also Max Shulman, *Rally Round the Flag, Boys!* (New York: Bantam, 1958), 24–26; Jerome Weidman, *The Enemy Camp* (New York: Random House, 1958), 23.

5. Harry Henderson, "How People Live in America's Newest Towns," part 1 of "The Mass-Produced Suburbs," *Harper's* (November 1953), 27; Herbert Gans, *The Levittowners* (New York: Pantheon, 1967), 22–23; William Whyte, *The Organization Man* (Garden City, N.Y.: Doubleday, 1956), 378.

6. Bennett Berger, *Working-Class Suburb: A Study of Auto Workers in Suburbia* (Berkeley: University of California Press, 1960); William Dobriner, *Class in Suburbia* (Englewood Cliffs, N.J.: Prentice-Hall, 1963), 97.

7. Robert Wood, *Suburbia: Its People and Their Politics* (Boston: Houghton Mifflin, 1958), 124; Schlivek, *Man in Metropolis,* 279–84; Dobriner, *Class in Suburbia,* 15, 96, 97, 106; Henderson, "How People Live," 27; Ralph Martin, "Life in the New Suburbia," *New York Times Magazine* (January 15, 1950); Greg Hise, "Home Building and Industrial Decentralization in Los Angeles: The Roots of the Postwar Urban Region," *Journal of Urban History* 19, no. 2. (February 1993), 95–125; Gans, *The Levittowners,* 22–23.

8. "The Insatiable Market for Housing," *Fortune* (February 1954); *Time* (July 3, 1950), 71; D. J. Waldie, *Holy Land: A Suburban Memoir* (New York: W. W. Norton, 1996), 34–35; Jackson, *Crabgrass Frontier,* 234–38.

9. William Whyte, "The Transients," *Fortune* (May 1953); William Whyte, "How the New Suburbia Socializes," *Fortune* (August 1953).

10. Lewis Mumford, *The City in History: Its Origins, Its Transformations, and Its Prospects* (New York: Harcourt, Brace, 1961), 486.

11. Jackson, *Crabgrass Frontier,* 213–15, 241.

12. Daniel Seligman, "The New Masses," *Fortune* (May 1959), 258; Harry Henderson, "Rugged American Collectivism," part 2 of "The Mass-Produced Suburbs," *Harper's* (December 1953), 82; Gibson Winter, *The Suburban Captivity of the Churches* (Garden City, N.Y.: Doubleday, 1961); "Is Our Religious Revival Real?" *McCall's* (June 1955); Whyte, *The Organization Man,* 405–22; Eric Goldman, "Good-by to the 'Fifties—and Good Riddance," *Harper's* (January 1960).

13. Whyte, "How the New Suburbia Socializes," 121.

14. Seligman, "The New Masses," 107; see also Phyllis McGinley, "Suburbia: Of Thee I Sing," *Harper's* (December 1949); Whyte, "How the New Suburbia Socializes," 121; Henderson, "How People Live."

15. E. Gartly Jaco and Ivan Belknap, "Is a New Family Form Emerging in the Urban Fringe?" *American Sociological Review* 18 (1953).

16. Margaret Mead, *And Keep Your Powder Dry* (New York: William Morrow, 1965), 270; Elaine Tyler May, *Homeward Bound: American Families in the Cold War Era* (New York: Basic Books, 1988), 11; *McCall's* cited in Alan Ehrenhalt, *The Lost City* (New York: Basic Books, 1995), 232.

17. Whyte, *The Organization Man,* 310.

18. Paul Glick, *American Families* (New York: Wiley and Sons, 1957), 104, 54; Joseph Veroff, Elizabeth Douvan, and Richard Kulka, *The Inner American: A Self-Portrait from 1957–1976* (New York: Basic Books, 1981), 140–93; Robert O. Blood and Donald M. Wolfe, *Husbands and Wives: The Dynamics of Married Living* (New York: Free Press, 1960).

19. Mary Cantwell, *Manhattan, When I Was Young* (New York: Penguin, 1996), 37.

20. Brett Harvey, *The Fifties: A Women's Oral History* (New York: HarperCollins, 1993), xiii; see also Wini Breines, *Young, White, and Miserable: Growing Up Female in the Fifties* (Boston: Beacon Press, 1992), 61; Agnes Rogers, "The Humble Female," *Harper's* (March 1950), 58; Barbara Ehrenreich, *The Hearts of Men: American Dreams and the Flight from Commitment* (Garden City, N.Y.: Anchor, 1983), 14.

21. Alfred Kinsey, *Sexual Behavior in the Human Female* (Philadelphia: W. B. Saunders, 1953), 331; quotes from Cantwell, *Manhattan*, 33, and Nora Johnson, "Sex and the College Girl," *Atlantic Monthly* (November 1959), 60. See also May, *Homeward Bound*, 114–34; John D'Emilio and Estelle Freedman, *Intimate Matters: A History of Sexuality in America* (New York: Harper and Row, 1988), 261–65; Harvey, *The Fifties*, 2–5.

22. Edwin Goldfield, *Statistical Abstract of the United States, 1965* (Washington, D.C.: U.S. Government Printing Office, 1965), 64; Ben Wattenberg, *This U.S.A.: An Unexpected Family Portrait of 194,067,296 Americans Drawn from the Census* (Garden City, N.Y.: Doubleday, 1965), 35; Harvey, *The Fifties*, 69–71; May, *Homeward Bound*, 6–8, 20, 137; D'Emilio and Freedman, *Intimate Matters*, 249; Sidonie Gruenberg, "Why They Are Marrying Younger," *New York Times Magazine* (January 30, 1955); Blood and Wolfe, *Husbands and Wives*, 122, 130; Veroff, Douvan, and Kulka, *Inner American*, 194–241; David Riesman, "Where Is the College Generation Headed?" *Atlantic Monthly* (April 1961); Mead, *Keep Your Powder Dry*, 270.

23. Goldfield, *Statistical Abstract*, 226, 228; Glick, *American Families*, 92–93; "The Rich Middle-Income Class," *Fortune* (May 1954), 95; Blood and Wolfe, *Husbands and Wives*, 99, 104–5; Daniel Bell, "The Great Back-to-Work Movement," *Fortune* (July 1956).

24. *Life* (June 27, 1955; and December 24, 1956); Katherine Hamill, "Working Wife, $96.30 a Week," *Fortune* (April 1953).

25. Quotes from, respectively, Marya Mannes, *But Will It Sell?* (New York: J. B. Lippincott, 1964), 34–35; Russell Lynes, "Teen-Agers Mirror Their Parents," *New York Times Magazine* (June 28, 1959), 22; Johnson, "Sex and the College Girl," 56.

26. Russell Baker, *The Good Times* (New York: William Morrow, 1989), 247.

27. Blood and Wolfe, *Husbands and Wives*, 11, 18–19.

28. *Life* (January 4, 1954; and July 16, 1956); see also Nancy Barr Mavity, "The Two-Income Family," *Harper's* (December 1951); Russell Lynes, *A Surfeit of Honey* (New York: Harper, 1957), 5, 57; Katherine Hamill, "Women as Bosses," *Fortune* (June 1956), 220; Mary Roche, "The New 'Servants': Machines and Husbands," *New York Times Magazine* (June 5, 1955); Clyde Kluckhohn, "Has There Been a Discernible Shift in American Values During the Past Generation?" in Elting Morison, ed., *The American Style* (New York: Harper and Row, 1958), 199; Mead, *Keep Your Powder Dry*, 270; Joanne Meyerowitz, ed., *Not June Cleaver: Women and Gender in Postwar America, 1945–1960* (Philadelphia: Temple University Press, 1994), 243; John Brooks, *The Great Leap* (New York: Harper and Row, 1966), 98.

29. *Life* (July 20, 1953); John Cheever, *The Housebreaker of Shady Hill and Other Stories* (New York: Harper and Brothers, 1958), 3.

30. Arnold Green, *Sociology: An Analysis of Life in Modern Society* (New York: McGraw-Hill, 1956), 370; Dobriner, *Class in Suburbia*, 9; William Dobriner, ed., *The Suburban Community* (New York: G. P. Putnam, 1958), 271–86; Gans, *The Levittowners*, 171–72; Henderson, "Rugged American Collectivism," 81; Henderson, "How People Live," 26; Morris Janowitz, *The Community Press in an Urban Setting* (Chicago: University of Chicago, 1967), 250–51.

31. First quote from Nora Sayre, *Previous Convictions: A Journey Through the 1950s* (New Brunswick: Rutgers University Press, 1995), 228–29; second quote from Mavity, "The Two-Income Family."

32. Mirra Komarovsky, *Blue-Collar Marriage* (New York: Random House, 1962), 51–52, 60; Dobriner, *Class in Suburbia,* 107; Raymond Chandler, *The Little Sister* (1949; reprint, New York: Vintage, 1988), 79.

33. Morison, *The American Style,* 199; William McIntyre, "War of the Sexes in Cartoonland," *New York Times Magazine* (September 1, 1957); *McCall's* cited in Ehrenhalt, *The Lost City,* 232.

34. "Upheaval in Home Goods," *Fortune* (March 1954); McPartland, *No Down Payment,* 82.

35. Quotes from "Upheaval in Home Goods"; see also Gilbert Burck and Sanford Parker, "The Wonderful, Ordinary Luxury Market," *Fortune* (December 1953); Gilbert Burck and Sanford Parker, "The Changing American Market," *Fortune* (August 1953); "Rich Middle-Income Class."

36. Thomas Hine, *Populuxe* (New York: Alfred A. Knopf, 1986), 12, 64; "Upheaval in Home Goods"; Emmet Hughes and Todd May, "Housing: The Stalled Revolution," *Fortune* (April 1957).

37. Hine, *Populuxe,* 20; quote from Margaret Halsey, *This Demi-Paradise* (New York: Simon and Schuster, 1960), 31–32.

38. "The Fabulous Market for Food," *Fortune* (October 1953), 135, 272.

39. Quotes from ibid., 138, 272; see also Roche, "The New 'Servants,'" 19; Bell, "Great Back-to-Work Movement."

40. Burck and Parker, "The Changing American Market"; William Whyte, "The Language of Advertising," *Fortune* (September 1952); for examples of the use of the word *modern* in ads, see *Saturday Evening Post* (February 10, 1951; March 31, 1951; May 5, 1951; and April 14, 1956).

41. Henderson, "How People Live," 28; Stan Opotowsky, *TV: The Big Picture* (New York: E. P. Dutton, 1961), 19–22; quote from Leo Bogart, *The Age of Television* (New York: Frederick Ungar, 1956), 68, 104.

42. Bogart, *The Age of Television,* 97, 150–52; John Houseman, "Battle over Television," *Harper's* (May 1950); Gilbert Seldes, "Can Hollywood Take Over Television?" *Atlantic Monthly* (October 1950); Robert Putnam, "The Strange Disappearance of Civic America," *PA* (Winter 1995).

43. Quote in Vance Packard, *The Hidden Persuaders* (Harmondsworth, Eng.: Penguin, 1962), 25; Daniel Seligman, "The Amazing Advertising Business," *Fortune* (September 1956); Gilbert Burck, "What Makes Women Buy?" *Fortune* (August 1956); Louis Kronenberger, *Company Manners: A Cultural Inquiry into American Life* (New York: Bobbs-Merrill, 1954), 25; Mannes, *But Will It Sell?;* Jules Henry, *Culture Against Man* (New York: Vintage, 1963), 45–99; Dwight Macdonald, *Against the American Grain* (New York: Random House, 1962); Norman Jacobs, *Culture for the Millions?* (Princeton, N.J.: D. Van Nostrand, 1961).

44. Charles Silberman, "Retailing: It's a New Ball Game," *Fortune* (August 1955).

45. Sidney Margolus, "Super Business of Supermarkets," *New York Times Magazine* (March 29, 1959), 24; Henderson, "How People Live," 27; Martin, "Life in the New Suburbia," 42.

46. George Sternlieb, *The Future of the Downtown Department Store* (Cambridge: MIT Press, 1962), 35, 64; Peter Muller, *Contemporary Suburban America* (Englewood Cliffs, N.J.: Prentice-Hall, 1981), 121–22; Jon Teaford, *The Twentieth-Century American City* (Baltimore: Johns Hopkins University Press, 1986), 105.

47. First quote from Margolus, "Super Business of Supermarkets," 24; second quote from C. B. Palmer, "The Shopping Center Goes to the Shopper," *New York Times Magazine* (November 29, 1953), 14.

48. Margolus, "Super Business of Supermarkets," 24, 31; Henderson, "How People Live," 31; Lee Rainwater et al., *Workingman's Wife* (New York: Oceana, 1959), 163–64; quote from Halsey, *This Demi-Paradise*, 2, 3–5.

49. John Keats, "Compulsive Suburbia," *Atlantic Monthly* (April 1960), 49.

50. Putnam, "Strange Disappearance"; Robert Putnam, "Bowling Alone: America's Declining Social Capital," *Journal of Democracy* 6, no. 1 (January 1995); Nicholas Lemann, "Naperville: Stressed Out in Suburbia," *Atlantic Monthly* (November 1989).

51. Whyte, *The Organization Man*, 310–29, 365–434; Henderson, "Rugged American Collectivism"; Gans, *The Levittowners*, 48, 51, 52, 59, 154, 193; Frederick Lewis Allen, "The Big Change in Suburbia," *Harper's* (June 1954); Frederick Lewis Allen, "Crisis in the Suburbs," *Harper's* (July 1954); Martin, "Life in the New Suburbia"; Janowitz, *Community Press*, 248–51.

52. Both quotes from Whyte, *The Organization Man*, 310, 317; see also Henderson, "How People Live," 28; Ehrenhalt, *The Lost City*, 210–20.

53. Whyte, *The Organization Man*, 378; Dobriner, *Class in Suburbia*, 92–93; see also Harvey, *The Fifties*, 71; Sidonie Gruenberg, "Homogenized Children of New Suburbia," *New York Times Magazine* (September 19, 1954); Gruenberg, "Why They Are Marrying Younger"; Dorothy Barclay, "Changing Ideals in Homemaking," *New York Times Magazine* (July 19, 1953); C. B. Palmer, "Spanking: The Pros, Cons, and Maybes," *New York Times Magazine* (March 7, 1954); Anne Kelley, "Suburbia: Is It a Child's Utopia?" *New York Times Magazine* (February 2, 1958); Eric Larrabee, *The Self-Conscious Society* (Garden City, N.Y.: Doubleday, 1960), 121; Henderson, "How People Live," 27; Dobriner, *The Suburban Community*, 157.

54. Quote from Henderson, "How People Live," 28; see also Berger, *Working-Class Suburb*, 69–71.

55. First quote from Harvey, *The Fifties*, 116; second quote from Henderson, "How People Live," 28.

56. Halsey, *This Demi-Paradise*, 37–38, 56; Time-Life Books, ed., *This Fabulous Century: 1950–1960* (New York: Time-Life Books, 1970), 167–72.

57. Daniel Seligman, "The Surge in School Building," *Fortune* (November 1958), 144; *Life* (September 26, 1955); Henderson, "Rugged American Collectivism," 81; Ehrenhalt, *The Lost City*, 215.

58. Frederick Lewis Allen, "The Spirit of the Times," *Harper's* (July 1952); see

also Frederick Lewis Allen, "The Unsystematic American System," *Harper's* (June 1952); Ehrenhalt, *The Lost City*, 216–20; Scott Donaldson, *The Suburban Myth* (New York: Columbia University Press, 1969), 155.

59. *Life* (December 26, 1955); Douglas T. Miller and Marion Nowak, *The Fifties: The Way We Really Were* (Garden City, N.Y.: Doubleday, 1977), 85–86; Will Herberg, *Protestant-Catholic-Jew: An Essay in American Religious Sociology* (Garden City, N.Y.: Doubleday, 1956), 14, 60–62; James Patterson, *Grand Expectations: The United States, 1945–1974* (New York: Oxford University Press, 1996), 329.

60. First quote from *Life* (April 6, 1953); second quote from Seligman, "The New Masses," 258; third quote from Henderson, "Rugged American Collectivism," 82; see also Winter, *Suburban Captivity*; "Is Our Religious Revival Real?"; Whyte, *The Organization Man*, 405–22; Goldman, "Good-by to the 'Fifties"; Ehrenhalt, *The Lost City*, 220–28; Herberg, *Protestant-Catholic-Jew*; John Mc-Greevy, *Parish Boundaries: The Catholic Encounter with Race in the Twentieth-Century Urban North* (Chicago: University of Chicago Press, 1996), 83–84.

61. McPartland, *No Down Payment*, 10–11.

62. Allen, "Spirit of the Times"; Allen, "Unsystematic American System"; Ehrenreich, *Hearts of Men*, 17; Henderson, "How People Live," 26; Henderson, "Rugged American Collectivism," 83, 86.

63. Shulman, *Rally!*, 33; see also Whyte, *The Organization Man*, 401; Gans, *The Levittowners*, 50–54; Henderson, "Rugged American Collectivism," 83; Russell Lynes, *The Domesticated Americans* (New York: Harper and Row, 1963), 282; Margaret Mead, *Life* (December 24, 1956); Agnes Meyer, "Women Aren't Men," *Atlantic Monthly* (August 1950), 33.

64. Gilbert Seldes, *The Great Audience* (New York: Viking, 1950), 244; see also Mead, *Life* (December 24, 1956); Mildred Adams, "Chip on Her Shoulder?" *Reporter* (July 4, 1950).

65. Mamie Eisenhower, cited in Meyerowitz, *Not June Cleaver*, 235; see also 232–33, 241, 249–50.

66. Time-Life Books, *This Fabulous Century*, 167–72; William Barry Furlong, "Big Strike: From 'Alley' to 'Supermarket,'" *New York Times Magazine* (November 29, 1959); Evan Jones, "Bowling Makes a Social Strike," *New York Times Magazine* (March 30, 1958); Putnam, "Bowling Alone."

67. Dobriner, *The Suburban Community*, 124; Janowitz, *Community Press*, 250–51; Berger, *Working-Class Suburb*, 72, 92, 116; Malcolm Gladwell, "True Colors: Hair Dye and the Hidden History of Postwar America," *New Yorker* (March 22, 1999), 73; McPartland, *No Down Payment*, 7.

68. Cara Greenberg, *Mid-Century Modern: Furniture of the 1950s* (New York: Harmony, 1984), 9, 35, 66–73; Whyte, *The Organization Man*, 350; Henderson, "How People Live," 26–27; "Upheaval in Home Goods"; Hine, *Populuxe*, 72.

69. Henderson, "How People Live," 26–27; Greenberg, *Mid-Century Modern*, 9.

70. Henderson, "How People Live," 29; "Lush New Suburban Market," 230; Lynes, *A Surfeit of Honey*, 66; Gans, *The Levittowners*, 48–49.

71. Fessenden Blanchard, "Revolution in Clothes," *Harper's* (March 1953); Dobriner, *The Suburban Community*, 152–53.

72. Henderson, "How People Live," 29, 31–32; see also Lynes, *A Surfeit of Honey,* 67–69.

73. Whyte, *The Organization Man,* 397.

74. Henderson, "How People Live," 27; Dobriner, *Class in Suburbia,* 107–8; Richard Yates, *Revolutionary Road* (Boston: Little, Brown, 1961), 59.

75. Henderson, "Rugged American Collectivism," 80; Peter Balakian, *Black Dog of Fate: A Memoir* (New York: Broadway Books, 1997), 37; Adlai Stevenson, "My Faith in Democratic Capitalism," *Fortune* (October 1955), 156; see also Seligman, "The New Masses"; McGinley, "Suburbia"; Donaldson, *The Suburban Myth,* 52–53, 108; Keats, "Compulsive Suburbia"; Whyte, *The Organization Man;* Alan Valentine, *The Age of Conformity* (Chicago: Regnery, 1954); C. Wright Mills, *White Collar* (New York: Oxford University Press, 1953), xviii; Robert Lindner, *Must You Conform?* (New York: Rinehart, 1956); Herbert Marcuse, *One Dimensional Man* (Boston: Beacon Press, 1964); Irving Howe, "This Age of Conformity," *Partisan Review* (January–February 1954); William S. White, "'Consensus American': A Portrait," *New York Times Magazine* (November 25, 1956).

76. Keats, "Compulsive Suburbia," 50.

77. *New York Times* cited in Whyte, *The Organization Man,* 346.

78. Veroff, Douvan, and Kulka, *Inner American,* xxxx; see also Blood and Wolfe, *Husbands and Wives,* 239–67.

79. Betty Friedan, *The Feminine Mystique* (New York: W. W. Norton, 1963), 15.

80. Goldfield, *Statistical Abstract,* 62; Wattenberg, *This U.S.A.,* 36; Blood and Wolfe, *Husbands and Wives,* 252; *Life* (December 24, 1956); see also Meyer, "Women Aren't Men"; Don Cortes, "What's Wrong with the American Woman?" *Atlantic Monthly* (August 1957); Johnson, "Sex and the College Girl"; Marya Mannes, "'Our Women Are Wonderful,' the American Said Sadly," *Reporter* (March 16, 1954); Mildred Adams, "What the Women's Vote Has Not Done," *New York Times Magazine* (August 20, 1950); Amaury De Riencourt, "Will Success Spoil American Women?" *New York Times Magazine* (November 10, 1957); Sloan Wilson, "The Woman in the Gray Flannel Suit," *New York Times Magazine* (January 15, 1956); Bernice Fitz-Gibbon, "Woman in the *Gay* Flannel Suit," *New York Times Magazine* (January 29, 1956); "That Woman in Gray Flannel: A Debate," *New York Times Magazine* (February 12, 1956); Bernice Fitz-Gibbon, "Tips for Would-be Women Bosses," *New York Times Magazine* (September 23, 1956); Marybeth Weinsten, "Woman's Case for Women's Superiority," *New York Times Magazine* (April 17, 1955); Will Chasen, "New York's Finest (Female Div.)," *New York Times Magazine* (November 20, 1955).

81. Quote from Harvey, *The Fifties,* 126–27; see also May, *Homeward Bound,* 22, 133, 183–207; "Crack-Ups in the Suburbs," *Cosmopolitan* (October 1960); Whyte, *The Organization Man,* 393–94; Martin, "Life in the New Suburbia," 42; Henderson, "How People Live," 27; Blood and Wolfe, *Husbands and Wives,* 239–67; Veroff, Douvan, and Kulka, *Inner American,* 140–93; William Whyte, "The Corporation and the Wife," *Fortune* (November 1951), 156, 158; William Whyte, "The Wives of Management," *Fortune* (October 1951), 87; Komarovsky, *Blue-Collar Marriage,* 60, 56–57.

82. Seligman, "The New Masses"; "Trouble in the Suburbs," *Saturday Evening Post* (September 17, 1955).

83. *Life* (October 18, 1954; and October 5, 1959).

84. Herbert Gold, "The Age of Happy Problems," *Atlantic Monthly* (March 1957); Macdonald, *Against the American Grain*, 59–61; Mannes, *But Will It Sell?*, 178.

85. Keats, "Compulsive Suburbia," 50; Harold Martin, "Are We Building a City 600 Miles Long?" *Saturday Evening Post* (January 2, 1960), 78, 79.

86. Donaldson, *The Suburban Myth*, 11–13, 155; Gans, *The Levittowners;* Dobriner, *The Suburban Community;* R. Martin, "Life in the New Suburbia"; Allen, "Big Change in Suburbia"; Henderson, "How People Live."

87. Dobriner, *The Suburban Community*, xvii; Edward Banfield and Morton Grodzins, *Government and Housing in Metropolitan Areas* (New York: McGraw-Hill, 1958), 18.

88. Dobriner, *Class in Suburbia*, 17–18, 110; Joshua Freeman, *Working Class New York* (New York: New Press, 2000), 172; Muller, *Contemporary Suburban America*, 121; "Lush New Suburban Market"; "Downtown Isn't Doomed!" *Saturday Evening Post* (June 5, 1954); Earle Schultz and Walter Simmons, *Offices in the Sky* (New York: Bobbs-Merrill, 1959), 228; C. T. Jonassen, *The Shopping Center versus Downtown* (Columbus: Ohio State University, 1955), 27–43; Economic Research Council, *Business Stability and Opportunities for Growth in the Syracuse Area* (Syracuse, N.Y.: Economic Research Council, 1954), 13; John Bodnar et al., *Lives of Their Own: Blacks, Italians, and Poles in Pittsburgh, 1900–1960* (Chicago: University of Illinois Press, 1982), 220; Berger, *Working-Class Suburb*, 72.

89. Dobriner, *Class in Suburbia*, 26, 17–18; Mumford, *City in History*, 486; Raymond Vernon, *The Changing Economic Function of the Central City* (New York: New York Metropolitan Region Study, 1959), 13–14, 50–51, 70–73, 74–79; Harold Martin, "Can We Halt the Chaos?" *Saturday Evening Post* (January 16, 1960); Teaford, *Twentieth-Century American City*, 98, 105, 111–13; Muller, *Contemporary Suburban America*, 7–8, 121, 122; Freeman, *Working Class New York*, 172; James Vance, *The Continuing City* (Baltimore: Johns Hopkins University Press, 1990), 489–502; Charles Silberman, "The Department Stores Are Waking Up," *Fortune* (July 1962).

90. Karl Marx, *Grundrisse* (New York: Vintage, 1973), 479.

Index